Building Futures

A reduction in the energy demand of buildings can make a major contribution to achieving national and international carbon reduction goals, in addition to addressing the interlinked issues of sustainable development, fuel poverty and fuel security. Despite improvements in thermal efficiency, the energy demand of buildings stubbornly remains unchanged, or is only declining slowly, due to the challenges posed by growing populations, the expectations of larger, more comfortable and better equipped living spaces, and an expanding commercial sector.

Building Futures offers an interdisciplinary approach to explore this lack of progress, combining technical and social insights into the challenges of designing, constructing and operating new low-energy buildings, as well as improving the existing, inefficient, building stock. The twin roles of energy efficiency, which is predominantly concerned with technological solutions, and energy conservation, which involves changing people's behaviour, are both explored. The book includes a broad geographical range and scale of case studies from the UK, Europe and further afield, including Passivhaus in Germany and the UK, Dongtan Eco City in China and retrofit houses in Denmark.

This book is a valuable resource for students and academics of environmental science and energy-based subjects, as well as construction and building management professionals.

Jane Powell is a Senior Lecturer in the School of Environmental Sciences at the University of East Anglia. She teaches and has research interests in the energy efficiency of buildings and low-carbon energy, in addition to environmental evaluation methodologies such as lifecycle assessment.

Jennifer Monahan is an Associate Tutor in the School of Environmental Sciences at the University of East Anglia. She has research interests in energy efficiency of buildings and low-carbon construction.

Chris Foulds is a Senior Research Fellow in the Consumption & Change theme of Anglia Ruskin University's Global Sustainability Institute. He is an environmental social scientist with a keen interest in the role of building occupants and professionals in the transition to a low-carbon building stock.

'This book is a must read for anyone planning a career in the built environment. It accurately captures the complexities of energy demand through focusing on interdisciplinarity and astutely assesses the opportunities of smart energy in homes, retrofits and new builds. Importantly, the book puts people right at the centre of its thinking.'

Philip Sellwood, Chief Executive, Energy Saving Trust

'We are on the threshold of a "next generation" of design thinking. This book explores some of the shortcomings in current approaches and provides blueprints for designing genuinely better buildings in which we can, over their and our lifetimes, successfully and comfortably survive in a warming and resource-challenged future.'

Sue Roaf, Heriot-Watt University, Edinburgh

Building Futures

Managing energy in the built environment

Jane Powell,
Jennifer Monahan
and Chris Foulds

Taylor & Francis Group

LONDON AND NEW YORK

from Routledge

First published 2016
by Routledge
2 Park Square, Milton Park, Abingdon, Oxon OX14 4RN

and by Routledge
711 Third Avenue, New York, NY 10017

Routledge is an imprint of the Taylor & Francis Group, an informa business

British Library Cataloguing-in-Publication Data
A catalogue record for this book is available from the British Library

Library of Congress Cataloging-in-Publication Data
Powell, Jane C., 1951– author.
  Building futures : managing energy in the built environment / Jane Powell,
Jennifer Monahan, Chris Foulds.
    pages cm
  Includes bibliographical references.
  1. Buildings—Energy conservation.   I. Monahan, Jennifer author.
  II. Foulds, Chris author.   III. Title.
  TJ163.5.B84P695 2016
  696—dc23   2015016788

ISBN: 978-0-415-72010-6 (hbk)
ISBN: 978-0-415-72012-0 (pbk)
ISBN: 978-1-315-67465-0 (ebk)

Typeset in Sabon
by Keystroke, Station Road, Codsall, Wolverhampton
Printed in Great Britain by Ashford Colour Press Ltd,
Gosport, Hants

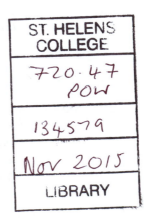

Contents

Illustrations

Figures

Tables

Foreword

Tackling the current, and unacceptable, levels of modern energy services deprivation (1.5 billion people without electricity and three billion without clean cooking) will have a transformative social impact through poverty alleviation and improved livelihoods, productivity, human health and energy security, together with environmental (e.g. indoor and outdoor air pollution, climate and ecosystem) co-benefits. That will, however, only be the case if there is a transition to a low-carbon economy and we decrease our reliance on fossil fuels, which is driving the single most serious environmental and development issue facing the world today – human-induced climate change.

There is no doubt that the composition of the atmosphere and the Earth's climate have changed since the Industrial Revolution, predominantly because of human activities. The atmospheric concentration of carbon dioxide has increased by over 30 per cent since the pre-industrial era, primarily from the combustion of fossil fuels and deforestation, accompanied by an increase in global mean surface temperature of about 0.85 °C. The IPCC *Fifth Assessment Report* projects that without additional actions to reduce greenhouse gas emissions the projected emissions growth is expected to result in global mean surface temperature increases relative to pre-industrial levels in 2100 from 2.5 °C to 7.8 °C when climate uncertainties are included. Even full implementation of the Cancun pledges will not set the world on a pathway to achieve the 2 °C goal. The impacts of climate change are likely to be extensive and primarily negative, and to cut across many sectors. For example, throughout the world, biodiversity at the genetic, species and landscape level is being lost, and ecosystems and their services are being degraded. In addition, food and water security are being threatened, as is human health, in many parts of the world.

Global warming caused by human-induced carbon dioxide increases is essentially irreversible on timescales of at least 1,000 years, mainly because of the ocean's storage of heat. Therefore, today's decisions about anthropogenic carbon dioxide emissions will determine the climate of the coming millennium. Without strong action to reduce emissions over the course of this century, we are likely to increase the atmospheric concentration by at least 300 ppm, taking concentrations to 750 ppm CO_2e or higher by the end of this century or the beginning of the next. The world's current commitments to reduce emissions are consistent with at least a 3 °C rise (50–50 chance) in temperature. Such a rise has not been seen on the planet for around three million years, much longer than *Homo sapiens* have existed. Human-induced carbon dioxide emissions pose a serious risk of a 5 °C increase to an average temperature not seen on the planet for 30 million years.

The major fossil fuel sources of carbon dioxide emissions are from energy production and use, with the key sectors being transportation, buildings and industry. This book, *Building Futures: Managing Energy in the Built Environment,* is essential reading for those interested in reducing emissions from buildings in a cost-effective and socially acceptable manner. It points out that energy is essential for all societies in order to meet their need for food, shelter and industry, and that the issue of energy use is not solely associated with climate change, but also energy security, fuel poverty, pollution, resource limitations and poverty alleviation. Universal access to clean energy services is vital for the poor, and a transition to a low-carbon economy will require rapid technological evolution in efficient energy use, environmentally sound low-carbon renewable energy sources and carbon capture and storage, coupled with an appropriate policy framework (including putting a price on carbon) and changes in behaviour. The longer we wait to transition to a low-carbon economy, the more we are locked into a high-carbon energy system with conse-quent environmental damage to ecological and socioeconomic systems. Therefore, if the world is to have any chance of achieving the politically stated goal of limiting the increase in global mean surface temperature to no more than 2 °C above pre-industrial levels (this would require a 50 per cent or greater reduction in global emissions by 2050), then energy demand in the building sector must decrease significantly in both existing and future buildings. This will be a major challenge for both developed and developing countries.

The book, which adopts an interdisciplinary approach to explore the key issues and solutions for reducing the energy demand of buildings in a warming world, discusses a wide range of issues, including approaches to reduce occupational energy use, energy embodied within buildings, the energy performance gap of buildings (i.e. the difference between the predicted and actual performance), approaches to retrofit existing buildings, energy efficiency building standards (specifically Passivhaus), building ventilation and building futures. Some of the key messages of the book include the following:

- Globally, energy demand continues to increase, with buildings currently being responsible for 34 per cent globally and 57 per cent in OECD countries.
- The number of households world-wide is predicted to grow by 67 per cent and the floor area of the service sector by almost 195 per cent by 2050.
- Addressing energy demand in the building environment is seen as critical by many countries trying to reduce their carbon emissions.
- Significant improvements in energy efficiency have driven down energy demand in all types of buildings, but this is counterbalanced by: demographic changes, such as increasing and ageing populations, leading to growing numbers of households; a desire for warmer and larger living and working spaces; and a greater number of appliances, particularly ICT equipment. In the service sector increases in energy demand are driven by more mechanical ventilation and air conditioning.
- The underlying reasons for energy use in buildings are complex and cross numerous disciplines, with no 'magic bullet' solutions. Even very technical aspects often have

social causes or consequences. Thus while substantial improvements to reduce energy consumption can be achieved with existing technologies, design strategies and technical innovations, they are insufficient alone to meet these challenges.

- To overcome the numerous barriers and market failures that undermine investment in energy efficiency in the built environment requires policy interventions; however, their effectiveness is highly dependent on how they are implemented and enforced.
- Given that 80 per cent of UK buildings in 2050 have already been built, it is imperative that the energy efficiency of the current stock is improved, while recognizing that the cost of retrofitting increases dramatically with the reduction in carbon emissions.
- Even newly constructed or retrofitted low-energy buildings often do not achieve the anticipated energy demand reductions.
- To reduce energy demand a broad interdisciplinary, systems-based approach is required that focuses not only on buildings during occupation but throughout their lifecycle, including the institutions that deliver them (planning, design, construction, policy).

The potential energy savings and consequent reduction in carbon dioxide emissions that can be achieved by the buildings sector are in line with the (UK) 80 per cent reductions that are required from the sector by 2050. As this book demonstrates, much of the technological knowledge is already available, but the application of that knowledge will not meet its target goals unless the users of those buildings are also given voice and the buildings are viewed as a socio-technological system. That is why I believe this book will be invaluable for all those involved in the building sector, from engineers to builders and occupiers.

Professor Sir Bob Watson CMG FRS
Monash Sustainability Institute, Monash University
& School of Environmental Sciences, University of East Anglia
February 2015

Professor Sir Bob Watson is Visiting Professor at Monash Sustainability Institute and Director of Strategic Development at the Tyndall Centre for Climate Change Research at the University of East Anglia. His many former roles include: DEFRA Chief Scientist; Chief Scientific Advisor and Director at the World Bank; Associate Director for Environment in the Office of Science and Technology Policy in the Executive Office of the President in the White House; and Director of the Science Division and Chief Scientist for the Office of Mission to Planet Earth at the National Aeronautics and Space Administration (NASA). He is also a Blue Planet Award Winner.

Preface

The built environment will form a key part of how we, as a society, plan to address key global challenges, such as climate change, fuel poverty and energy security. These challenges emphasise the urgency of calls to reduce the energy demand of our building stock. But how do we practically go about reducing demand? What are the main obstacles to or enablers of change? What are we to realistically expect if we continue to focus on the technological 'solutions' that currently dominate policy? How can we best measure and monitor progress and performance? How could building energy demand change in the future? And, ultimately, what does all this mean for our future aspirations of constructing a lower-energy building stock?

In exploring such questions, we adopt an interdisciplinary approach that predominantly uses technical issues as starting points, before utilising social science perspectives for additional insight. Interdisciplinarity is important because although many challenges to reducing energy demand appear to be technical, in reality they are more complex and require the input of less technical disciplines such as the social sciences. Using this approach can provide a clearer understanding about how exactly those technologies will be adopted and used in reality.

This desire to write an interdisciplinary book on the built environment caused us problems when liaising with various publishers. We discovered that the vast majority are ill-equipped to even review an interdisciplinary book proposal, as editorial departments tend to be organised along traditional disciplinary boundaries (e.g. engineering, sociology). It is thus unsurprising that there are so few books that adopt an interdisciplinary perspective – spanning both technical and social disciplines – as we do here. It is exactly for these reasons that we are grateful to Routledge, who were able to engage with such interdisciplinarity.

This book consists of eight chapters. The first two chapters – *Energy in the built environment* (Chapter 1) and *Reducing energy demand* (Chapter 2) set the scene and provide context and a solid rationale for the rest of the book. Specifically, the first chapter starts with a review of global energy demand and the challenges that demand is causing, before focusing on how energy is used in buildings and what is driving changes in that demand. The second chapter introduces different strategies for lowering the energy demand of buildings, which assists in detailing some of the more technical concepts that other chapters build upon and essentially take for granted. The following chapters then explore in greater depth: how one takes into consideration the energy demand of buildings across their whole lifecycle, including that which is embodied into materials (Chapter 3); differences between the expected energy performance and what is actually achieved (Chapter 4); the need to retrofit our existing building stock (Chapter 5); the Passivhaus energy efficiency standard (Chapter 6); and the challenges of meeting the ventilation requirements of our building stock (Chapter 7). Chapter 8 reflects more broadly on the cross-cutting synergies and challenges from the previous chapters and looks at possible ways forward.

This book is intended for those from both interdisciplinary and single-discipline backgrounds, who have an interest in learning more about how energy is used at present and is likely to be used in the future, as part of designing, constructing, retrofitting and occupying the international building stock. The book is inspired by an array of disciplines, including architecture, building engineering, economics, environmental sciences, human geography, sociology and science and technology studies. Specifically, this book is targeted at undergraduate- and Masters-level students, while PhD students will hopefully find it provides useful background to their research. We also hope this book will provide insight for construction, architecture, planning and building management professionals, particularly through the inclusion of real-world examples, practical guidance and a jargon-free narrative.

The title of this book rather presumptuously refers to the 'built environment', but we actually write about 'buildings'. In terms of energy and demand, buildings are the central component of the built environment, but we recognise the importance of considering how the future of our buildings is managed in the context of the wider built environment. While energy management is often used interchangeably with the targeting of energy savings, in this book we implicitly adopt a broader approach. Indeed, we emphasise how different ways (e.g. design, planning, construction, maintenance, occupation) of engaging with energy-demanding building technologies represent some form of energy management.

This book is also firmly about the 'environment'. We are not engineers; we are not qualified to write about how to construct a building. Rather, we are environmental scientists, for whom trying to understand and manage environmental impacts and challenges is central to our focus and beliefs. We were also, at different times, all environmental science students at the University of East Anglia, who are interdisciplinary to the core. Our joint experience of teaching an energy module from an interdisciplinary perspective, and trying to find an appropriate interdisciplinary textbook, inspired the writing of this book.

We hope this book will contribute to an improved understanding of the complexity of reducing energy demand in buildings and of showing that there are no easy answers or quick-fix solutions. Nevertheless, we hope a holistic, interdisciplinary approach to this problem will lead towards sustainable ways forward for society.

Jane Powell, Jennifer Monahan, Chris Foulds
March 2015

Acknowledgements

We are especially grateful to the following colleagues who have kindly offered comments and insights on draft book chapters:

Carrie Behar, Energy Institute, University College London.

Dr María Dolores Bovea, Department of Mechanical Engineering and Construction, Universitat Jaume I, Castellon, Spain.

Peter Chisnall, The Energy Practice Ltd, Colchester; Green Peaches C.I.C., Colchester.

Professor Douglas Crawford-Brown, Cambridge Centre for Climate Change Mitigation Research (4CMR), University of Cambridge.

Dr Mahoo Eftekhari, School of Civil and Building Engineering, Loughborough University.

Jo Franklin, Independent Consultant, Norfolk.

Martin Ingham, Linktreat Ltd, Norwich.

Patricia Kermeci, School of Environmental Sciences, University of East Anglia.

Dr Jason Palmer, Cambridge Architectural Research, Cambridge.

Alison Pooley, Department of Engineering and the Built Environment, Anglia Ruskin University.

Dr Fausto Sanna, Department of Architecture, Arts University, Bournemouth.

Dr Henrik Schoenefeldt, Kent School of Architecture, University of Kent.

Nicola Terry, Director of Qeng Ho Ltd, Cambridge.

Professor Andrew Watkinson, School of Environmental Sciences, University of East Anglia.

We also thank our colleagues at Routledge who patiently guided us first-time authors through the intricacies of publication.

Acronyms and units

ach	air changes per hour
ASHP	air source heat pump
BEMS	building energy management systems
BIM	buildings information modelling
BPIE	Building Performance Institute Europe
BRE	Building Research Establishment
BREEAM	Building Research Establishment Environmental Assessment Methodology
BSI	Building Standard Institute
BSRIA	Building Services Research and Information Association
C	carbon
CAD	computer-aided design
CEN	Comité Européen de Normalisation
CFL	compact fluorescent light bulb
CHP	combined heat and power
CIBSE	Chartered Institute of Building Services Engineers
CfSH	Code for Sustainable Homes
CO_2	carbon dioxide
CO_2e	carbon dioxide equivalent
CoP	coefficient of performance
COPD	chronic obstructive pulmonary disease
DEC	Display Energy Certificate
DECC	(UK) Department for Energy and Climate Change
DOE	(US) Department of Energy
EC	European Commission
ECO	Energy Company Obligation
EIO	economic input–output (analysis)
EEA	European Economic Association
EIA	(US) Energy Information Administration

ELCD	European Reference Life Cycle Database
EPBD	EU Directive on the Energy Performance of Buildings
EPC	Energy Performance Certificate
EPD	Environmental Product Declaration
EST	Energy Saving Trust
EU	European Union
GHG	greenhouse gas
GJ	gigajoule(s)
GSHP	ground source heat pumps
GW	gigawatt(s)
GWh	gigawatt-hour(s)
GWP	global warming potential
ha	hectare
HFC	hydrofluorocarbon(s)
HVAC	heating, ventilation and air conditioning
ICE	Bath University's Inventory of Carbon and Energy
ICT	information and communications technology
IEA	International Energy Agency
IPCC	Intergovernmental Panel on Climate Change
ISO	International Organization for Standardization
J	joule(s)
kg	kilogram(s)
$kgCO_2$	kilogram(s) of carbon dioxide
$kgCO_2e$	kilogram(s) of carbon dioxide equivalent
kW	kilowatt(s)
kWh	kilowatt-hour(s)
$kWh/(m^2 \cdot a)$	kilowatt-hour(s) per m^2 per year
kWp	kilowatt-peak
K	kelvin
LCA	lifecycle assessment
LCD	liquid crystal display
LED	light emitting diodes
LEED	Leadership in Energy and Environmental Design
m	metre(s)
$m^3/(h \cdot m^2)$	air permeability: cubic metres of air per hour per square metre of material
MEV	mechanical extract ventilation
MJ	megajoule(s)
$mtCO_2$	million tonne(s) of carbon dioxide
$mtCO_2e$	million tonne(s) of carbon dioxide equivalent
mtoe	million tonne(s) of oil equivalent
MW	megawatt(s)
MWh	megawatt-hour(s)
MVHR	mechanical ventilation with heat recovery
OECD	Organisation for Economic Co-operation and Development
Pa	pascal(s)
PAS 2050	Publically Available Specification 2050

PHI	Passivhaus Institute
PHPP	Passivhaus Planning Package
POE	post-occupancy evaluation
ppm	parts per million
PROBE	Post Occupancy Review of Buildings and their Engineering
PSD	passive solar design
PV	solar photovoltaics
RIBA	Royal Institute of British Architects
SAP	Standard Assessment Procedure
SBEM	Simplified Building Energy Model
SBS	sick building syndrome
SHW	solar hot water
t	tonne
TPES	total primary energy supply
TSB	Technology Strategy Board (now Innovate UK)
TW	terawatt(s)
TWh	terawatt-hour(s)
UN	United Nations
UNEP	United Nations Environment Programme
UNFCCC	United Nations Framework Convention on Climate Change
VOC	volatile organic compound
W	watt(s)
WAP	Weatherization Assistance Program
$W/(m^2 \cdot K)$	U-Value: watts per metre squared per kelvin
Wh	watt-hour(s)
WHO	World Health Organization

1 Energy in the built environment

1.1 Introduction

Energy is essential for all societies in order to meet their needs for food, shelter, industry and commerce. Energy-dense fossil fuels have allowed most countries to develop in a way that would have amazed pre-industrial society. Escalating populations, with rising standards of living, drive an ever-increasing global demand for energy, resulting in the interlinked problems of climate change, pollution, limits to resources, fuel poverty and energy security. As energy resources become scarcer, their extraction will be more expensive and cause increased environmental damage, leading to higher prices and increased fuel poverty.

However, despite this need to curb our demand for energy, in the words of the International Energy Agency, 'taking all new developments and policies into account, the world is still failing to put the global energy system onto a more sustainable path' (IEA, 2012, p. 1). Not only are we failing to progress sufficiently towards our long-term carbon targets, but we are also failing to tackle the inequity in access to energy resources, the health issues associated with energy use and the limits to our energy resources.

Globally we mainly use energy for the residential sector (25 per cent), services (9 per cent), industry (31 per cent) and transport (31 per cent) (2012 figures) (Figure 1.1). In addition to the residential sector, the energy used by 'services', which includes commercial, retail, public services, leisure sector, hospitals and education sectors, is also almost entirely associated with buildings. World-wide, buildings are considered to have an energy demand of over one-third of global energy and half of global electricity demand (IEA, 2013). So, if we are to tackle our energy-related problems successfully, the built environment needs to be a major contributor to reducing energy demand and carbon emissions. Technically, this reduction is relatively straightforward for buildings, compared to the logistical problems posed by other sectors such as transport. However, improvements in the energy efficiency of buildings are undermined by various factors, such as population growth and increases in living standards. Indeed, the number of households world-wide is

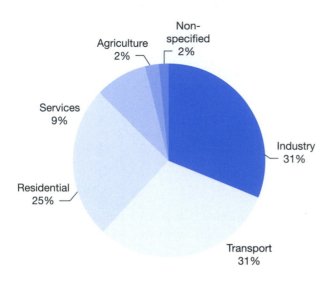

Figure 1.1
Global energy demand by sector (%) (2012).

Source: IEA (2014).

predicted to grow by 67 per cent and the floor area of the service sector by almost 195 per cent by 2050 (from 2007) (OECD/IEA, 2011).

Significant progress has been made to improve the thermal envelope of buildings, but the actual energy demand has not declined as much as predicted. In addition to technical failings, this problem is also due to how our buildings are used, suggesting that a more holistic, interdisciplinary approach is needed to explore the technical, environmental and social challenges of designing and using low-energy buildings. We are unlikely to meet our energy and carbon targets unless an interdisciplinary approach is used to tackle this challenge. The need for such an approach has been echoed by the UK Royal Academy of Engineers (2010), which identified an urgent need for multidisciplinary research in building design, engineering, energy and carbon efficiency that would inform the construction industry of the potential of alternative initiatives.

This first chapter initially explores why energy demand is a problem, discussing climate change, fuel poverty and energy security. We then examine how much energy we use globally, in the EU and the UK, and then focus on different energy sectors (industry, commercial, residential and transport), identifying why buildings are particularly important. This will include why the amount and types of energy we use are important, and in particular the carbon consequences of the fuel used, especially the electricity fuel mix. Trends in energy demand will be examined, identifying the underlying demographic and sociotechnical drivers for changes in energy use in buildings. We then go on to discuss the challenges to reducing energy demand in buildings and explore examples of policy solutions that are being used to address these challenges.

1.2 Why is energy use a problem?

Energy, of course, is used by everyone, in all parts of the global economy, although its form varies considerably. The poorest people use 'local fuels' such as wood and manure to cook basic ingredients and to keep warm, while Western societies use energy, predominantly from fossil fuels, to extract resources and manufacture products, for transportation and to provide services (heat, cooling and hot water) at work and in our homes. But this increasing demand for energy is leading to many important global challenges, in particular climate change, fuel poverty and energy security.

1.2.1 Climate change

> Warming of the climate system is unequivocal, and since the 1950s, many of the observed changes are unprecedented over decades to millennia. . . . It is extremely likely that human influence has been the dominant cause of the observed warming since the mid-20th century.
>
> (IPCC, 2013, p. 2)

Human activity, resulting in the emission of carbon dioxide (CO_2) from the combustion of fossil fuels, is augmenting the atmospheric concentration of greenhouse gases (GHGs) at an unprecedented rate. The accumulation of atmospheric concentrations of all GHGs has increased since the inception of the Industrial Revolution in 1750 (IPCC, 2013), with global CO_2 emissions from fossil fuel combustion and cement production increasing on average by 2.5 per cent per year over the past decade (Friedlingstein, 2014). The impacts on human society will be widespread and will involve destructive weather events, the disruption of food production and impacts on human health (IPCC, 2014a). While there is still debate over the potential future impacts of climate warming, a 2 °C rise has become the accepted threshold beyond which climate change effects will be 'dangerous' (Copenhagen Accord, 2009).

However, there is a 'significant gap' between the current trajectory of global GHG emissions and the 'likely chance of holding the increase in global average temperature below 2 °C or 1.5 °C above pre-industrial levels' (UNFCCC, 2011, quoted in Peters *et al.*, 2012, p. 1). To keep temperature increases below 2 °C is likely to require challenging global CO_2 mitigation rates of over 5 per cent per year, partly because of the likelihood of increased emissions in some regions, in addition to meeting the long-term carbon quota (Raupach *et al.*, 2014). Moreover, if there is a delay in starting mitigation activities, the target of remaining below 2 °C becomes even more difficult, if not unfeasible. Many of these higher rates of mitigation rely on negative emissions, using emerging technologies such as carbon capture and storage possibly linked to bioenergy, which have high risks due to potential delays or failure in their development and large-scale deployment (Peters *et al.*, 2012). As a consequence of the challenges of achieving these mitigation rates, the likelihood of keeping climate change within 2 °C is being increasingly called into question, with a trajectory of 4, 5 or even 6 °C seeming more likely (Jordan *et al.*, 2013).

Nevertheless, society's principal response to the challenge of climate warming needs to be one of mitigation, with urgent and radical moves towards decarbonisation (Anderson *et al.*, 2008). The global political response towards achieving long-term reductions in emissions began with the United Nations Framework Convention on Climate Change in 1994 and the first legally binding protocol, the Kyoto Protocol, adopted in 1997 and entered into force in 2005. Under the Kyoto Protocol, governments committed to reduce their country's emissions by a target date. The UK, for example, committed to reduce its emissions by 12.5 per cent from 1990 levels by 2012; this it easily achieved, mainly as a result of the economic recession and exporting its carbon footprint (i.e. many products used in the UK are manufactured in other countries, so the associated carbon emissions are not attributed to the UK).

Globally, carbon emissions nearly doubled from 1973 (15,633 $mtCO_2$) to 2012 (31,734 $mtCO_2$) (OECD/IEA, 2014). Carbon dioxide emissions from global fossil fuel combustion and cement production have increased every decade over the last 50 years, starting from an average 3.1 ± 0.2 GtC/yr in the 1960s, to 8.6 ± 0.4 GtC/yr during 2003–2012 (Le Quéré *et al.*, 2014). (Note that these values are expressed as carbon rather than carbon dioxide.) It might have been expected that the global financial crisis in 2008–2009, with its impact on economic activities, would have slowed the rate of growth in emissions. Its effects were short-lived, however, owing to a substantial growth in the emerging economies, a return to emissions growth in developed economies and an increase in the fossil-fuel intensity of the world economy (Peters *et al.*, 2012). In terms of CO_2 emissions per person, the CO_2 emissions in 2012 were 1.4 tC per capita/yr globally, and 4.4 (USA), 1.9 (China), 1.9 (EU) and 0.5 (India) tC per capita/yr (Le Quéré *et al.*, 2014).

Carbon dioxide emissions from fossil fuel combustion and industrial processes contributed to approximately 78 per cent of the total GHG emissions during the period 2000–2010, while the use of buildings contributed 19 per cent of GHG emissions (2010) (IPCC, 2014b). Note that the emissions associated with the construction of buildings is included in the industrial sector. Even though many countries are seeking ways to meet carbon emission targets by reducing their energy demand, the energy demand from buildings is projected to double and CO_2 emissions to increase by 50 to 150 per cent by 2050 in baseline scenarios. However, there are considerable opportunities to reduce this demand using energy-efficient technologies, policies and know-how. In developed countries it is also considered that up to 50 per cent of energy demand could be reduced through behavioural and lifestyle changes by 2050 (IPCC, 2014b).

1.2.2 *Fuel poverty*

Fuel poverty is a term used in the UK and some other EU countries to describe the lack of access to energy. In the United States 'energy insecurity' is used, and in 20 OECD countries 'lacking affordable warmth' (Liddell and Morris, 2010). There is also conflicting use of the terms 'energy' and 'fuel poverty' within the EU (Thomson and Snell, 2013a), with 'energy poverty' being the more common term in developing countries. All these terms, however, focus on a lack of access to adequate, 'clean' (i.e. with low particulate emissions) energy, whether this is due to (1) energy resources not being available, (2) the lack of an adequate energy infrastructure or (3) the inability to pay for energy. In this book we will use the term 'fuel poverty'.

It has been estimated that 2.7 billion of the poorest people (40 per cent of the global population) rely on traditional biomass (e.g. manure, wood, crop wastes) and simple stoves or open fires for cooking and heating their homes; 1.2 billion (17 per cent) light their homes with simple kerosene lamps; and 1.3 billion (18 per cent of the global population) lack access to electricity (IEA, 2014; WHO, 2014). There is clear evidence that indoor air pollution from open fires and stoves in the home increases the risk of chronic obstructive pulmonary disease, acute respiratory infections and many other illnesses, especially in women and children. It has also been estimated that indoor air pollution results in 4.3 million premature deaths per year from respiratory and cardiovascular diseases, together with cancer (WHO, 2014). In addition, open fires and unsafe stoves and lighting can lead to significant numbers of accidents.

The Millennium Development Goals (UN Millennium Project, 2005), a blueprint agreed to by all countries, does not refer directly to energy access, although energy is a fundamental prerequisite for achieving many of the goals, as social and economic development is unlikely to occur without access to reliable and affordable energy services. If we are to alleviate poverty and hunger, families need affordable clean and/or more efficient fuels for cooking staple foods, lighting, food preservation and irrigation. Adequate health care requires transport, lighting and refrigeration. If women (and children) are freed from having to spend hours each day collecting wood they will have new opportunities for education and enterprise.

With the poorest half of the world's households relying on solid fuel (coal, biomass), one effective method of averting millions of premature deaths and significantly reducing carbon emissions is the promotion of low-emission stove technologies. It has been calculated (Wilkinson *et al.*, 2009) that the distribution in India of 15 million improved cooking stoves per year for ten years, which would reduce the proportion of traditional stoves used in homes from 68 to 13 per cent, would reduce the total number of premature deaths of 240,000 young children from acute lower respiratory infections, and 1.8 million adult deaths from ischaemic heart disease and chronic obstructive pulmonary disease (COPD). In addition 0.1–0.2 $mtCO_2e$ per million people could be avoided in one year (Wilkinson *et al.*, 2009).

In developed countries, energy-inefficient dwellings, low household income and increases in the cost of energy all contribute to increasing fuel poverty. Most EU countries have low levels of awareness of fuel poverty (Thomson and Snell, 2013a), despite the fact that 9.8 per cent of EU-27 households and 15.8 per cent of the 12 new Member States' households could not afford to heat their homes sufficiently in 2011 (Thomson and Snell, 2013b). Only three EU countries – the UK, Republic of Ireland and France – have official definitions for fuel poverty. In England a household is considered to be in fuel poverty if their fuel costs are above the national average and, if having spent this amount, their remaining income leaves them below the official poverty line (DECC, 2014a). Most other countries do not recognise the concept of fuel poverty, even though it is likely many of their population are affected by it. Interestingly, it has been shown that many southern, as well as eastern, European countries suffer more from fuel poverty than more northerly, colder countries (Thompson and Snell, 2013a). Cold-related deaths are higher in Spain and Portugal than in Scandinavia (Healy, 2003).

Excess winter deaths, the increase in deaths in the winter months compared to the summer months, are usually caused by respiratory and cardiovascular illnesses during normal winter temperatures, when the outdoor temperature drops below 5–8 °C (Public Health England, 2013). These illnesses are considered to be a consequence of cold homes, with the death rate rising by about 2.8 per cent for every 1 °C drop in the external tempera-ture for people living in the coldest 10 per cent of homes in the UK (Wilkinson *et al.*, 2011). The 25,100 (moving five-year average) excess winter deaths per year in the UK (ONS, 2014) are regarded as a serious public health issue. The World Health Organization (WHO) recommends a minimum temperature of 21 °C in living rooms and 18 °C in other rooms (WHO, 2007).

1.2.3 Energy security

The security of the energy supply is central to energy policy-making. However, its focus has changed significantly over the last few decades. Historically, energy security focused on the finite stock of fossil fuels and their use within a centralised energy model. The use of indigenous energy resources was consequently given priority in national energy policies (Chalvatzis, 2012). However, this has changed in many countries in order to accommodate decentralised renewable energy generation, and with the move from national energy independence to the diversity of energy supply (Flavin and Dunn, 1999). A diverse supply is considered to be more resilient to physical and political disruption and to spread the risks. Energy security, therefore, now tends to be discussed in terms of diversified supplies and trade routes rather than energy independence (Nuttall, 2010). The World Bank Group (2005) refers to energy security as an approach that must focus on three different pillars: (1) energy efficiency, (2) diversification of energy supplies and (3) energy price volatility. It considers that a country's energy security means sustainable production and use of energy at affordable prices. More recently energy security has been defined in terms of five dimensions: availability, affordability, technology development, sustainability and regulation (Sovacool and Mukherjee, 2011).

For many countries, energy security is considered to be low due to a reliance on a few energy resources such as coal and gas, which are often imported. The expansion of low-carbon energy has the benefits of not only lowering carbon emissions by changing the fuel mix, but also increasing the diversity of supply and exploiting indigenous renewable energy resources. Although this book is concerned with energy demand rather than supply, the fuel mix, particularly the electricity fuel mix, of a country has a significant influence on the carbon emissions of buildings. Countries that generate substantial proportions of their electricity from renewable resources or nuclear have low carbon emission factors for their grid electricity. Compare, for example, the UK emission factor (0.5085 kgCO$_2$/kWh) with Norway (0.0022 kgCO$_2$/kWh), which has considerable hydroelectricity capacity, and France (0.0709 kgCO$_2$/kWh), which generates 75 per cent of its electricity from nuclear (Brander *et al.*, 2011). For countries with high emission factors, the benefits from reducing electricity demand are considerable. There is also a stronger case for investment in low-carbon micro-generation to reduce the demand for grid electricity.

In the UK, as well as incentivising renewable electricity, the production of heat from renewable resources, such as timber and solar, has been encouraged for homes and other buildings by the introduction of the Renewable Heat Incentive (DECC, 2011).

1.3 How much energy do we use?

If we reduce the amount of energy we use, decarbonise our supply and manage what we produce more efficiently, we should be able to reduce carbon emissions, fuel poverty and use of non-renewable resources. So why isn't this happening on a sufficiently fast timescale? To examine this question, we first need to understand our current energy use and past trends. For instance, how much energy are we using, and is this increasing or decreasing? What types of energy are being used and what for? What are the trends and why are they occurring?

1.3.1 Global energy supply and demand

Energy is measured in different ways. The primary production of energy is defined as 'any extraction of energy products in a useable form from natural sources' (Eurostat, 2013), such as the mining of coal, oil extraction or generation of electricity from wind or nuclear (primary electricity). Delivered energy or energy demand is the energy delivered to the consumer's door, excluding energy used, for example, for transformation purposes such as electricity generation from coal.

Total primary energy supply (TPES) is the energy used within a country, excluding international marine and aviation bunkers. 'Bunkers' refer to energy delivered to shipping and aviation engaged in international aviation and navigation (IEA, 2012). TPES is measured in terms of million tonnes of oil equivalent (mtoe).

The global energy supply (TPES) has more than doubled in 37 years (6,107 mtoe in 1973 to 12,717 mtoe in 2010) at a rate of 2.9 per cent/year. In 2010, 88 per cent of the global TPES was from fossil fuels, with oil being the largest amount followed by coal (Figure 1.2). In OECD countries, the TPES increased at the lower rate of 1.1 per cent/year with a total increase of 41.8 per cent from 1973 to 2011 (3,740 mtoe in 1973 to 5,305 mtoe in 2011) (IEA, 2012).

It should be noted that global energy statistics are likely to provide an underestimate of energy use because they only include energy resources that have been commercially exchanged. Energy that is 'free', such as wood and manure that is collected and used by a significant proportion of rural populations, is not accounted for.

Looking to the future, the IEA (2012) provides estimates for two scenarios: first, the New Policies Scenario (NPS), which is based on policy commitments and plans that have already been announced; and, second, the 450 Scenario (450S), which is based on policies under consideration. The NPS estimates a TPES increase of 33 per cent (1.33 per cent/year) by 2035, whereas the 450S estimates a lower increase of 17 per cent (0.67 per cent/year) (IEA, 2012) (Figure 1.3). A comparison of the TPES by fuel, for 2010 and the two scenarios, indicates an increase in the demand for all fuels under the NPS, while the 450S reduces demand for oil and coal/peat, increasing the proportional use of other fuels.

Although it is interesting to examine energy in terms of primary production, it is perhaps more meaningful, in a book concerned with sectoral (e.g. residential, transport) energy demand to explore energy in terms of demand (i.e. in residential terms, energy delivered to one's home). The main difference is that energy 'lost' in generation, transmission and

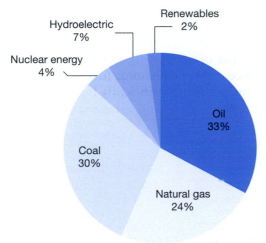

Figure 1.2
Global primary energy consumption by fuel type (%) (2013).

Source: BP (2014).

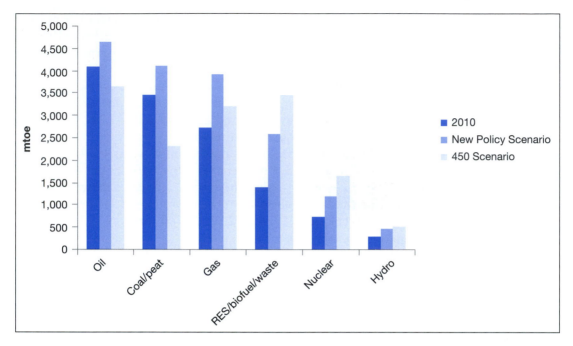

Figure 1.3 Total primary energy supply (TPES) by fuel type in 2010 and EIA scenarios (NPS and 450S) in 2035 (mtoe) (RES: renewable energy sources).

Source: IEA (2012b).

distribution processes is excluded (Figure 1.4). The unit usually used to describe these values varies across the different energy databases, but is generally termed (total) final energy demand or consumption. Strictly speaking, energy cannot be 'consumed', but this term is widely used in energy statistics; we generally use the term 'energy demand' in this book, or occasionally 'final energy demand' if this is more appropriate.

Global energy demand is 8,639 mtoe/year (2012), with the OECD responsible for 42 per cent of that demand (Figure 1.5). It is interesting to note that OECD energy demand has declined since 2007, levelling off in recent years (Figure 1.6), while that of China has increased 2.5 times since 1990, and the remaining Asian countries has doubled. Although energy demand in Africa is very small, it has also increased substantially (1.8 times) since 1990.

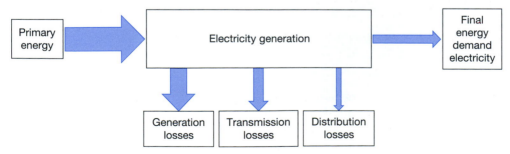

Figure 1.4 A simple flow chart of primary energy to final energy demand for electricity.

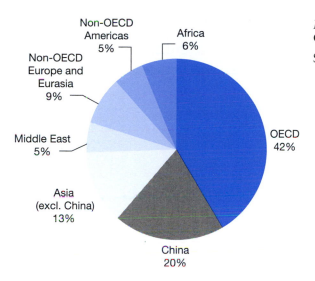

Figure 1.5
Global energy demand by region (%) (2012).

Source: IEA (2014).

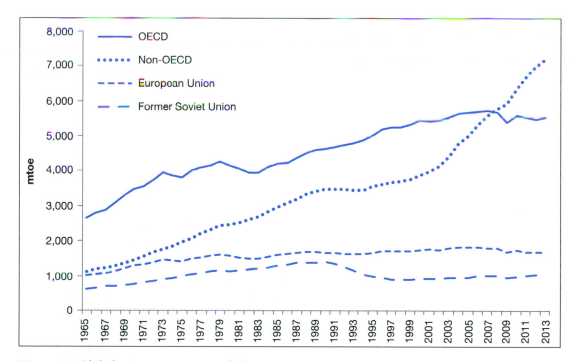

Figure 1.6 Global primary energy supply by region (mtoe) (1965–2013).
Source: BP (2014).

The sector with the largest energy demand is transport, both globally and in the EU, but if 'residential' and 'services' are added together to create a 'buildings' sector, this is the largest at 34 per cent globally (Figure 1.1) and 40 per cent in the EU (Figure 1.7). Globally, industry and transport have similar sized energy demands to each other, whereas transport is responsible for a larger proportion of energy demand than industry in the EU.

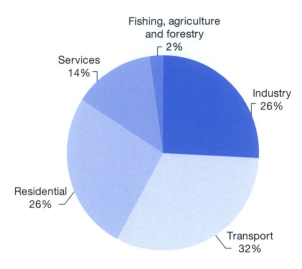

Figure 1.7
Final energy demand by sector in EU-28 countries (%) (2012).

Source: Eurostat (2014b).

In the emerging economies, particularly the BRICS countries (Brazil, Russia, India, China and South Africa), rapid urbanisation is leading to substantial construction. Indeed, in China more new houses were constructed in 2010 than the entire housing stock of Spain (Economist Intelligence Unit, 2012), although in absolute terms this might be misleading given that China's housing stock was already considerably larger than Spain's. Expansion of the building sector of this magnitude will result in significant increases in energy demand of the construction industry, as well as the residential and service sector during occupation.

1.3.2 European energy demand

The 28 countries of the EU account for approximately 13 per cent of the global energy demand (1,153 mtoe of energy in 2010) (Eurostat, 2014a). Overall, this energy demand increased slightly (2.3 per cent) between 1990 and 2012, but in recent years (2005–2012) decreased by 7.1 per cent (EEA, 2015a). However, the majority of this recent decrease is likely to be predominantly as a result of the economic downturn in 2008, rather than structural shifts in energy demand activities, such as reductions in manufacturing or increases in the service sector.

The total energy demand in the EU-28 countries is 26 per cent for residential buildings and 14 per cent for services (Figure 1.7). Household energy demand has increased by approximately 12 per cent from 1990 to 1996, but since then has been relatively constant, only fluctuating with the weather (Eurostat, 2014b) (Figure 1.8). Recently, on a per person basis, the final household energy demand in the EU-27 decreased by 0.6 per cent (2005–2010) (EEA, 2012b). However, this relatively constant overall energy demand hides underlying changes that are discussed further below.

The fastest-growing sector in terms of energy demand is 'services', which increased by 37.3 per cent from 1990 to 2012, followed by transport, which increased by 25.0 per cent (Figure 1.8). Although there have been improvements in transport fuel efficiency, these have been largely offset by increases in the distances travelled. This relates to increases in the ownership of private cars, particularly in the new EU Member States, and changes in lifestyle. There have also been significant and rapid increases (73 per cent 1990–2009)

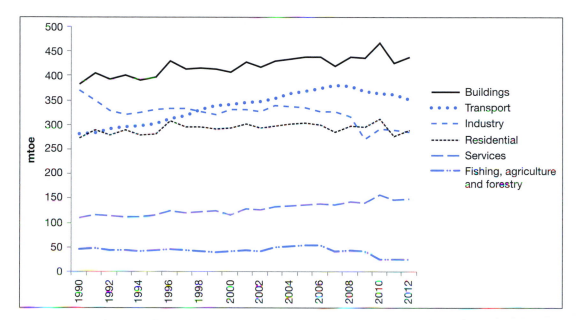

Figure 1.8 Final energy demand in EU-28 countries (mtoe) (1990–2012).
Source: Eurostat (2014b).

in national and international passenger aviation (EEA, 2015a). Transport energy overtook industrial energy demand, the sector with the previously highest energy demand, in 1998.

The use of energy by industry has declined mainly due to the shift of a significant proportion of heavy industry to countries such as India and China, and more recently to the economic recession, which has affected all sectors. To some extent the industrial energy decline has also coincided with an increase in the energy demand in the service sector, although at a smaller scale.

The main energy resource used in EU residential buildings is gas (39 per cent), followed by electricity (25 per cent), while in non-residential buildings 48 per cent of the energy used is electricity (Table 1.1). In EU households, 67.1 per cent of energy demand is for space heating, with almost 80 per cent of individual heating systems being gas-fired. The demand for gas has increased by 42 per cent for households (1990–2012) (EEA, 2015a), but this is mainly due to the increase in the number of households, as on an individual household basis the energy for heating has declined. The demand for electricity in the EU has increased over the last 20 years; 38 per cent for residential buildings and 74 per cent for non-residential buildings (BPIE, 2011). This is mainly due to increases in the number of appliances. The drivers for increases in electricity are discussed in Section 1.4.4.

Over the last 20 years (1990–2010) the energy efficiency of EU homes has improved by 18.7 per cent (Odyssee, 2014), mainly due to improvements in the thermal efficiency of buildings and a significant increase in the adoption of high-efficiency boilers. New-build homes in the EU-27 (2011) use, on average, 40 per cent less energy than homes built in 1990 (EEA, 2012a). Although this improved energy efficiency has led to a reduction in energy used for space heating on a per dwelling (Figure 1.9) and per floor area (m^2) basis, an increase in the floor area, both per person and per home, plus an increase in the number

Table 1.1 Energy fuel mix for energy demand in non-residential buildings and residential buildings in different regions of Europe (%)

Fuel	Residential energy (%)			Non-residential energy (%)
	Central and East	North and West	South	
Gas	26	41	39	29
Electricity	16	26	29	48
Oil	3	19	19	15
DH/CHP	13	5	0	6
Solid fuels	14	1	0	1
Renewable energy systems	21	9	12	1

Source: BPIE (2011).

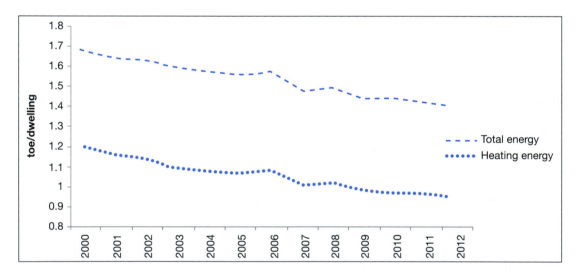

Figure 1.9 Total energy and heating energy per household in EU-28 countries + Norway (toe/dwelling) (1990–2012).

Source: Odyssee (2014).

of households, has resulted in only a slight reduction in the total energy used for space heating of housing. It is estimated that 20 per cent of improvements in the energy demand for heating has been offset by the increase in the size of homes (Enerdata, 2012). Demographic changes are discussed further in Section 1.4.1.

There are significant differences in the amount of energy used per person in different countries for space heating and cooling, hot water, cooking and appliances, ranging from less than 0.2 toe per person in Malta to more than 0.8 toe per person in Finland (2010) (EEA, 2012b). These differences are influenced by many factors, including climate, the

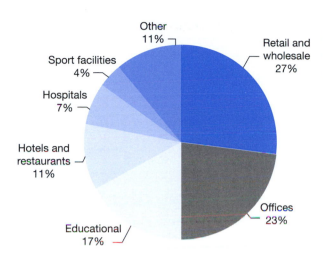

Figure 1.10
Proportions of non-residential building types in Europe (%) (2011).

Source: BPIE (2011).

thermal efficiency of homes, types of heating system and energy prices, as well as different consumption patterns (EEA, 2012b) due to sociocultural differences in how people live their lives and thereby demand energy.

1.3.3 Building stock in the EU

The energy demand of a region or country is strongly influenced by the physical characteristics of its building stock. In the EU, 75 per cent of the building stock (in terms of floor space) consists of residential buildings, of which 64 per cent are single-family houses and 34 per cent apartments (BPIE, 2011). Although there are many different types of homes, the non-residential building sector is even more complex and heterogeneous, in that it includes retail buildings (from small shops to supermarkets), hospitals, hotels, restaurants, offices, schools, universities and sports centres. Based on floor area, retail and wholesale buildings form the largest proportion of non-residential buildings followed by offices (Figure 1.10) (BPIE, 2011).

1.3.4 UK energy demand

The total UK energy demand is 142.4 mtoe (2013), which in 2012 was approximately 12 per cent of the EU-28 energy demand. Transport energy, which currently demands the largest proportion of energy (37.6 per cent) has increased rapidly over the last 33 years (1980–2013), overtaking industrial and residential energy demand in the 1980s (Figure 1.11). As in the rest of Europe, UK industrial energy demand has steadily declined with the reduction in heavy industry and increased energy efficiency. The energy demand of buildings (residential plus services), the majority of which is used for heating, varies according to the weather. Residential energy demand during this period was highest in the cold year of 2010, but then fell dramatically in 2011 due to a mild winter.

On a temperature-corrected basis (Figure 1.12), residential energy demand has gradually declined since 2004 (note shorter timescale), seemingly reflecting improvements in energy efficiency. However, there are also other underlying drivers for energy demand, discussed

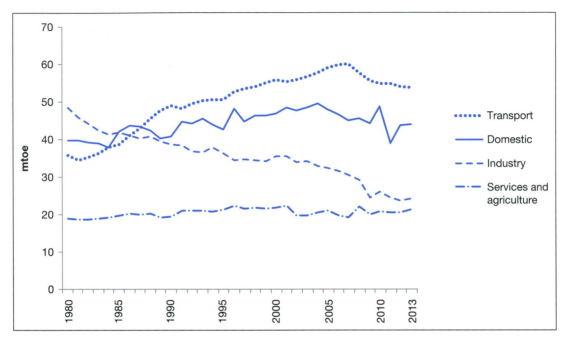

Figure 1.11 UK energy demand (mtoe) (1980–2013).
Source: DECC (2014b).

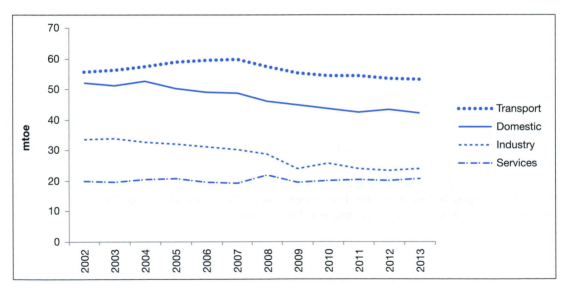

Figure 1.12 UK energy demand by sector corrected for temperature (mtoe) (2002–2013).
Source: DECC (2014c).

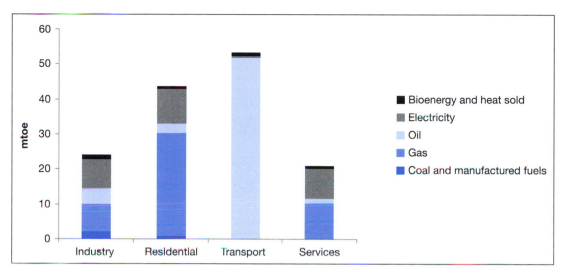

Figure 1.13 UK energy demand by sector and fuel (mtoe) (2013).
Source: DECC (2014d).

in Section 1.4. The energy demand from services has remained fairly constant, varying slightly with changes in the economic situation.

Considered as a whole, the main fuel used in the UK is oil (42.4 per cent of total demand), but this is the dominant fuel only for transport and is just a small proportion of the other sectors' fuel mix (Figure 1.13). In contrast, gas and electricity dominate in the sectors where buildings are important. Gas (32.6 per cent of total demand) is the predominant residential fuel and is also used in industry and services, while similar amounts of electricity are used in industry, services and residential sectors (overall 19.2 per cent of total demand).

In the UK, 49.1 per cent (67.2 mtoe) of total energy demand (2013) is from buildings; this includes space heating within industrial buildings. The majority of the energy demand is for space heating (62.4 per cent), but lighting and appliances (17.7 per cent) and water heating (14.1 per cent) also form significant proportions (Figure 1.14). Three-quarters of space heating is fuelled by gas, which is also the most used fuel overall (60.5 per cent), with electricity making up the majority of the remainder (28.3 per cent).

While the energy demand for space heating has increased since 1980, mainly due to demographic changes, the demand for hot water and cooking has declined (Figure 1.15). The drivers behind these changes are discussed in Section 1.4.

1.4 Drivers of energy demand in the built environment

Despite decades of policies to reduce energy demand, it has steadily increased in the EU and globally. In Europe, higher personal incomes are thought to be the main, long-term driver of increased energy demand, which in turn lead to higher standards of living, changing how we live, work and play, and thus ultimately demand energy (e.g. higher levels of thermal comfort, increased ownership of domestic appliances) (EEA, 2015b). Changes in technology, especially digital technology, have also led to increases in energy demand.

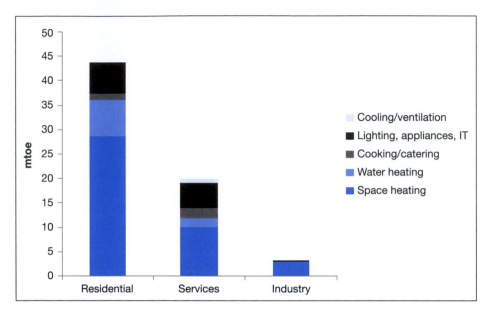

Figure 1.14 UK building-related energy demand in the residential, service and industrial sectors (mtoe) (2013).

Source: DECC (2014e).

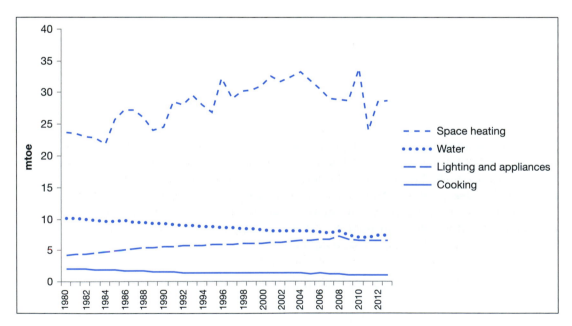

Figure 1.15 UK residential energy demand by end use (mtoe) (1980–2013).

Source: DECC (2014c).

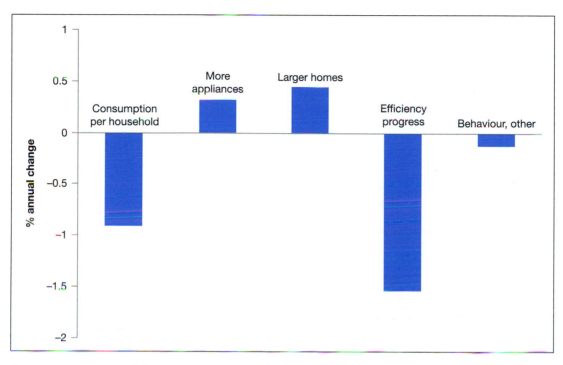

Figure 1.16 Drivers of change in EU-28 countries, annual average energy demand per household (% annual change) (1990–2012).

Source: EEA (2015b).

There are many complex and often contradictory drivers for energy demand in buildings (Figure 1.16). Improvements in thermal efficiency, such as more efficient new homes and the retrofit of older homes, have played a substantial role. Of the efficiency savings from residential space heating, one-third are considered to have been driven by new, more stringent, building regulations. For example, a newly (2015) built home will use 40 per cent less energy than one built in 1990. However, efficiency improvements, including more efficient heating appliances, are thought to be counterbalanced to some extent by demographic changes such as increases in total population, ageing populations, smaller households and larger homes. Other drivers include the substantial increase in the number and usage of appliances and lighting. Of the energy efficiency improvements in homes achieved through technological development, half are thought to have been offset by increases in the size of homes and increasing numbers of appliances (EEA, 2015b). These are discussed in more detail in the following sub-sections.

1.4.1 Demographic drivers

An increase in the global population is a particularly powerful driver of increasing energy demand. The United Nations' population projections indicate that there will be considerable population growth in developing countries, particularly in urban areas (Figure 1.17). In Europe, low birth rates plus a growing life expectancy are resulting in shrinking populations

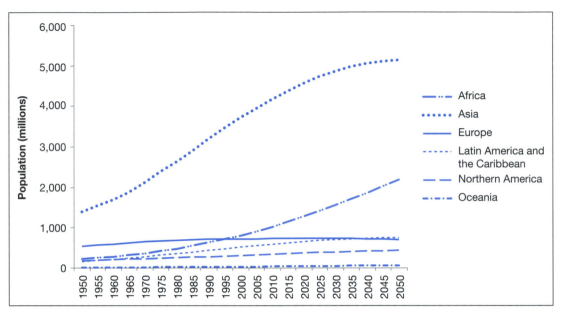

Figure 1.17 Total world population by area (millions) (1950–2050).

Source: UN (2012).

(York, 2007), although the UK population, in contrast, is increasing, partly due to net immigration (Figure 1.18).

Many countries such as Japan and Italy have ageing populations. Although ageing is not so advanced in developing countries, the populations of some countries are ageing at a faster rate than developed countries, meaning that ageing populations will become a significant problem for them in the future (UN, 2013). Globally, the proportion of people aged 60+ is predicted to increase from 12 per cent in 2014 to 21 per cent by 2050. In Europe, 23 per cent of people are already aged 60+ and that proportion is expected to increase to 34 per cent by 2050. In Eastern Asia, the currently lower proportion of people aged 60+ (14 per cent in 2009) is predicted to increase to 42 per cent by 2050 (UN, 2014).

The size and age structure of a population, together with the household size (i.e. number of people) and level of urbanisation, all have a significant influence on energy demand. One reason is that older people tend to live in smaller households, which usually use considerably more energy per person than larger households (Jiang and Hardee, 2011). A country with an ageing population consequently has increasing numbers of these high energy demanding homes. In the EU, the trend towards smaller households is predicted to continue until 2020, when it will stabilise. However, in developing countries this trend could continue to 2050 or 2060, mainly as a result of average household sizes currently being much larger and having the potential to reduce considerably in size (Jiang and O'Neill, 2009).

In the UK, where the number of households is increasing at a rate of 0.88 per cent per year (Palmer and Cooper, 2013a), it is estimated that a one-person household uses approximately 60 per cent more energy per person than a two-person household (Boardman *et al.*, 2005), with residential energy demand increasing approximately in line with increasing household numbers (Shorrock and Utley, 2003). It has also been estimated that

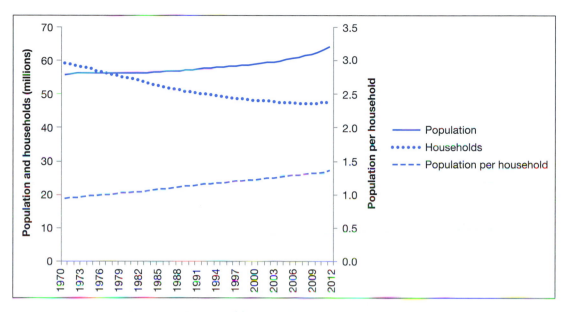

Figure 1.18 UK population (millions) and household size (1970–2012).

Source: Palmer and Cooper (2013b).

on average a one-person household consumes 38 per cent more products, 42 per cent more packaging and 55 per cent more electricity per person than a four-person household (Williams, 2007).

In developed countries, there has been a notable increase in dwelling space per person. The average area of a dwelling in the EU-15 increased from 86 to 92 m² (1990–2009), while the household size decreased from 2.8 to 2.4, resulting in a 20 per cent increase in floor area per person (EEA, 2012b). This is likely to lead to a higher energy demand for heating and cooling, as well as for construction materials, unless these trends are offset by rising energy and material efficiency.

1.4.2 Thermal efficiency of homes

There have been considerable improvements in the thermal efficiency of homes, particularly in the cooler developed countries, mainly due to increased loft and cavity wall insulation and the introduction of double glazing, resulting in reduced heat loss (Figures 1.19, 1.20). In UK homes, the introduction of more energy-efficient condensing and combination (usually gas) boilers has also led to reduced energy demand (Figure 1.21). Combination, or 'combi', boilers heat water for both central heating and residential hot water, negating the need for the storage of hot water. They are, however, now being phased out in favour of condensing combi boilers. Modern condensing boilers, which extract heat from their waste gases, can obtain 90 per cent efficiency. In the UK, gas condensing boilers became compulsory for new homes in 2005 and for oil condensing boilers in 2007. The replacement of existing boilers is being funded under the Energy Company Obligation (ECO) and to a lesser extent the Green Deal. Around 79 per cent of UK boilers are now condensing or combi-condensing (Palmer and Cooper, 2013b).

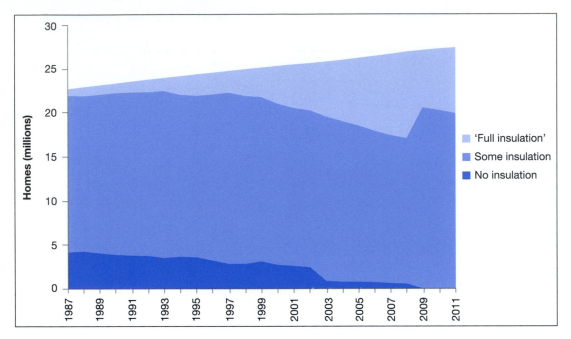

Figure 1.19 Number of UK homes with full, some or no insulation (1987–2011).

Source: Palmer and Cooper (2013b).
Full insulation is defined as ≥ 100 mm loft insulation, plus cavity wall insulation, plus ≥ 80 per cent of rooms with double glazing. Households with no insulation have none of these technologies. Remaining households are classified as having some insulation. There is a discontinuity in the data between 2008 and 2009 due to a change in the statistical methodology used.

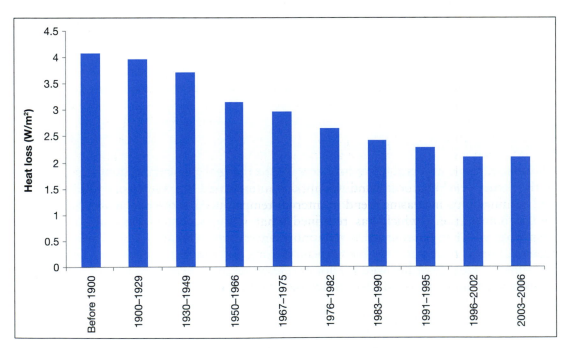

Figure 1.20 Heat loss of UK housing by year of construction (W/m^2).

Source: Palmer and Cooper (2013b).
Heat loss is calculated from the heat loss through the thermal envelope of a building and through ventilation divided by the total floor area.

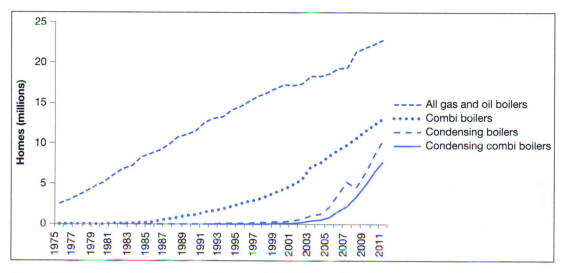

Figure 1.21 Number of UK homes with gas and oil boilers and with energy-efficient boilers (millions). Source: Palmer and Cooper (2013b).

Under the EU Ecodesign Directive, new efficiency standards for central heating boilers are being introduced that are predicted to reduce the EU's energy demand by 10 per cent by 2020. A new ecolabelling scheme will also help to make clear the lifetime energy savings potential of the boilers. Under this scheme a condensing boiler will be reduced to an A category and an A+++ category will be created for super-efficient boilers, such as heat pumps.

1.4.3 Internal temperatures

Another trend that has led to increased energy demand is an increase in average internal temperatures of residential (Figure 1.22) and non-residential buildings. In the UK, average internal temperatures have risen by 5.6 °C (1970–2011). How homes are heated has changed dramatically during this period, from primarily heating just one room (e.g. the lounge), often with a solid fuel fire and no heating for the bedrooms, to primarily using a gas-fired central heating system to heat every room in the home. Indeed, in the 1970s it was not uncommon for bedroom windows in the winter to have a layer of ice on the inside in the morning. This increasing trend in internal temperatures represents a shift in thermal comfort conventions, which has redefined what is deemed to be 'normal' residential temperatures. The consequence is that more energy needs to be used to provide these higher (normal) temperatures. The 'flattening out' of the temperatures in the last 13 years suggests an optimal temperature for comfort may have been reached. However, it should be noted that these values are modelled rather than measured.

1.4.4 Electricity and appliances

The demand for electricity in homes in Europe has recently increased by 35.9 per cent in residential buildings (1990–2012) (Eurostat, 2014b) (Figure 1.23) and 77.8 per cent in

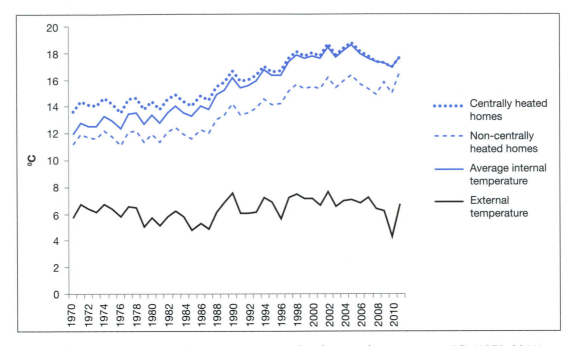

Figure 1.22 Average UK whole-house winter internal and external temperatures (°C) (1970–2011).
Source: Palmer and Cooper (2013b).

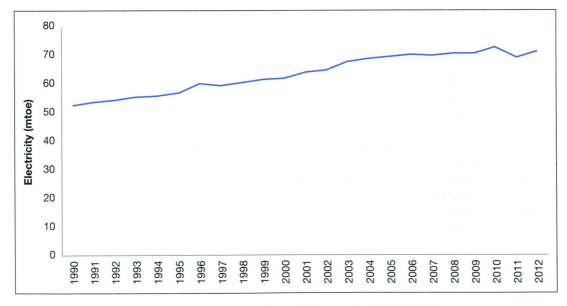

Figure 1.23 Total electricity demand in EU-28 households (mtoe) (1990–2012).
Source: Eurostat (2014b).

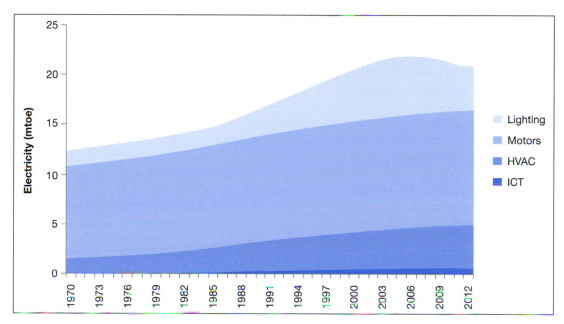

Figure 1.24 UK service sector non-residential appliances electricity demand (mtoe) (1970–2013).
Source: DECC (2014f).
Motors: electric motors and fans; HVAC: heating, ventilation and air conditioning.

non-residential buildings (1990–2009) (BPIE, 2011). This is mainly due to an increase in the use of electrical appliances, in particular consumer electronics such as televisions and DVDs, and information and communication technology (ICT) equipment (e.g. laptops and mobile phones) (plus the increase in the number of households). On a per person basis, recent household electricity demand has increased by 4.6 per cent (2000–2012) in the EU-28 plus Norway, but with increases of more than 20 per cent in Bulgaria, Estonia, Latvia, Lithuania, Romania and Turkey (EEA, 2012b) due to increases in affluence in these countries.

In the service sector, high energy consuming technologies such as electric motors dominate electricity demand (Figure 1.24).

Although the majority of electricity demand in homes is for appliances, 21.7 per cent of residential electricity is used for space heating and cooling (JRC/IE, 2009). The electricity demand of air cooling systems overall in Europe is only 4.4 per cent of total electricity, but in warmer Mediterranean countries it is more significant. In particular, in warmer countries there has been a rapid increase in the purchase of small residential air conditioners, the extensive use of which during the summer months has been one of the main drivers to increases in electricity demand and power peaks (JRC/IE, 2009).

In homes, the demand for consumer electronics (e.g. televisions) and ICT (e.g. computers) is predicted to continue to increase unless more ambitious policy measures are introduced to increase energy efficiency (IEA, 2012) or to reduce usage. However, the energy demand from lighting and cold appliances, such as fridges, is declining due to improvements in energy efficiency, and especially the widespread introduction of low-energy light bulbs.

The increase in the ownership of appliances is clearly demonstrated in the UK, particularly the increase in consumer electronics and ICT appliances. The number of televisions increased from 17 million in 1970 to nearly 62 million in 2011. Laptops, which first appeared in UK homes in 1991, are now estimated to number nearly 21 million (DECC, 2012). These increases are partly due to increases in the number of households and higher disposable incomes.

As a consequence of this increase in appliances, the electricity demand of household appliances increased by 45 per cent in total, or 1.1 per cent per year (1980–2013). Consumer electronics, currently 26 per cent of appliance electricity demand, have increased particularly rapidly, by 4.4 per cent per year (Figure 1.25). The greatest increase, however, is in the electricity demand of ICT which, following the mass production and rapid society-wide uptake of home computers, has increased from 0.18 to 6.71 TWh (1980–2013), a staggering 83 per cent per year. Cold appliances (e.g. freezers) also increased, but their electricity demand then declined from the late 1980s, reflecting improvements in their energy efficiency. The introduction of energy-efficient lighting has also resulted in a reduction in electricity demand in the service sector (Figure 1.24), as well as in homes (Figure 1.25) (DECC, 2014c).

There has also been an increase in the standby electricity used by ICT and consumer electronics. Standby power refers to the power used by an appliance or device while it is switched off but is still using power. Standby electricity was thought to be around 6 per cent of all household electricity demand (JRC/IE, 2009), but a recent UK survey found standby electricity amounted to 9–16 per cent of total electricity demand (EST, 2012).

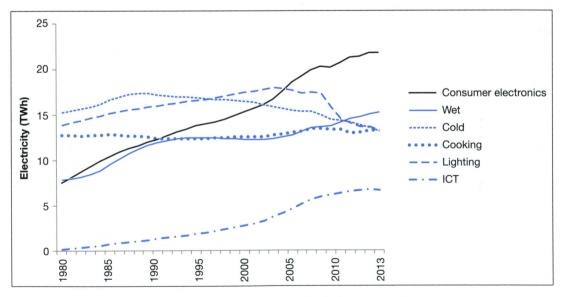

Figure 1.25 UK household appliance electricity demand (TWh) (1980–2013).

Source: DECC (2014c).
Consumer electronics includes: televisions, DVDs, HiFi equipment; ICT includes: desktop and laptop computers, modems, tablets.

1.5 Challenges to reducing energy demand in buildings

As we have seen in the previous sections, the changing energy demand of buildings is complicated in that it is influenced by a range of drivers, such as increases in population and declining household sizes, as well as improvements in energy efficiency and living standards. Overall these drivers have not resulted in significant reductions in energy demand despite pressing environmental and social challenges. Why is this, and why is it proving so difficult to reduce the energy demand in buildings? The tools and technologies to produce energy-efficient buildings, described in Chapter 2, are well known, so why aren't low-energy buildings the norm? It seems that this is partly because the production of new buildings, let alone energy-efficient ones, is a complex process, not only in the design and construction process but also in the institutional arrangements that determine the governance of the building industry as a whole (Metz *et al.*, 2007).

Investment in energy efficiency in the built environment is considered to be undermined by a number of challenges (IEA, 2008), the most commonly identified relating to finance, market failure and institutional organisation. This section introduces these challenges, which are discussed in more detail throughout the rest of the book.

1.5.1 Financial barriers

One of the most commonly identified challenges to reducing energy demand in buildings is considered to be cost (SBCI, 2007). The specifying and purchasing of better-performing insulation or more efficient equipment and appliances often involves higher upfront capital costs, which the developer or owner may not be able to afford because of limited capital, for instance, despite the benefits of lower energy costs in the future. There are also hidden costs associated with adopting new technologies (Carbon Trust, 2005). These include real or perceived risks (such as poor performance), additional costs of implementation (such as training, certification and liabilities) and transaction costs of obtaining the knowledge and information necessary for making decisions and sourcing the chosen option. For example, switching to high-efficiency boilers requires time to research the different technological options, make an investment decision and source the new technology, which may not be readily available on the market. The new boiler may not be compatible with existing fittings, and require dedicated fittings that are a further additional cost and require expertise to install. These hidden costs can be high in the built environment due to its fragmented and diverse nature. The absorption of some of these hidden costs by the contractor may be standard practice, although some of these costs may be short term and transitional while a new technology becomes the norm.

For further discussion of the costs of retrofit, see Section 5.4.1.

1.5.2 Market failures

Market failures have been identified as a key barrier to energy efficiency in the built environment (Carbon Trust, 2005). A market failure is where a competitive and freely functioning market does not deliver an efficient allocation of resources. It occurs where market structures and constraints in the market prevent the trade-off between energy reduction investment and energy-saving benefits (Metz *et al.*, 2007). According to traditional economics, market failure may happen where there is imperfect information and knowledge, for instance, about a new low-carbon technology that is incomplete, unavailable, expensive and/or difficult to obtain. Market failure can also be a product of regulatory challenges (e.g. planning policy preventing incorporation of renewable energy systems), institutional challenges (e.g. the fragmented nature of the design build process), lack of materiality (i.e. the relatively low cost of energy compared to other costs) and where there are split incentives (e.g. the landlord has the burden of investment in energy efficiency while the tenant accrues the benefit through reduced running costs – see Section 5.4.6).

1.5.3 Institutional failures

Typically the process of constructing a building is considered linear and sequential. Consequently, it may appear to be straightforward, but the process is also split into many constituent parts, with each decision involving a multitude of actors affecting design, construction and operation. In addition, the process is also highly fragmented, requiring the input of many specialised actors, each making decisions specific to a relatively small part of the build process. As each actor may be working to optimise their own particular and individual part, this can result in specific areas being energy efficient, but not the whole building itself. As a consequence, this fragmentation and diversity in the building process can lead to sub-optimal outcomes (Metz *et al.*, 2007). Achieving a low-energy building requires a whole-building approach that considers built form, environmental factors, envelope and mechanical and electrical techniques as part of an integrated system. This is discussed further in Chapter 6.

1.6 Policy solutions to reducing energy demand in buildings

Policy solutions are considered essential to overcoming the key challenges introduced in Section 1.5, and driving forward energy efficiency in the built environment. More than 30 policy instruments, used to reduce the energy demand from buildings, have been identified and aggregated into four categories of mainstream policy (SBCI, 2007) (Table 1.2).

Instruments for control and regulation refer to laws or regulatory requirements that mandate the improvement of energy efficiency in devices, practices or systems (Vreuls, 2005). These are found most commonly in the form of building codes, building regulations and minimum energy performance standards (Table 1.2). Building codes or regulations that include energy performance standards prescribe minimum acceptable energy performance standards for buildings and services. For example, the EU Directive on the Energy Performance of Buildings (EPBD), in force since January 2003, requires all EU Member States to have minimum energy performance requirements for new buildings and those undergoing significant refurbishment.

Minimum energy performance standards are concerned with appliances and devices such as domestic appliances, electronic equipment, office equipment, electrical motors and heating, ventilation and air conditioners (HVAC) equipment. Typically, performance standards have two parts. The first states the minimum performance standard of a specific device type, while the second specifies testing procedures required to estimate or classify the energy efficiency of the devices in question (Vreuls, 2005).

Table 1.2 Mainstream policy instrument categories and instruments for energy efficiency in the built environment

Category	Measure
Control and regulation	Building codes and enforcement Appliance standards (minimum energy performance) Energy-efficiency obligations and quotas Procurement regulations
Information	General information projects Information centres Product labelling and certification Energy audits Education and training Demonstration projects Governing by example
Economic	Cooperative procurement (bulk purchasing) schemes Capital subsidies, grants or subsidised loans Taxes, tax exemption, tax credits Energy efficiency certificate schemes Energy and carbon trading (Kyoto mechanisms)
Voluntary agreements	Voluntary certification and labelling Voluntary negotiated minimum energy standards and agreements Public leadership programmes Awareness raising, education and information campaigns Detailed billing and disclosure programmes
Combinations of policy measures	

Source: adapted from Vreuls (2005) and UNEP (2007).

The effectiveness of regulation and control has been demonstrated to be highly dependent upon enforcement (Ürge-Vorsatz *et al.*, 2003). Without enforcement, sanctions or other mechanisms for compelling compliance, there is a risk that regulations will not achieve their goal. Furthermore, control and regulatory instruments also require monitoring, evaluation and regular revision to respond to technological developments and changes in the markets as they respond to policy.

Information-based instruments aim to persuade consumers to change their behaviour by increasing consumer awareness and understanding of energy-efficient products, services and behaviours and their economic and environmental benefits (Koeppel and Ürge-Vorsatz, 2007). Information, support and voluntary measures include product labelling, energy audits, education and training, information centres, demonstration projects and media campaigns (Table 1.2) (Vreuls, 2005). These measures are often used to complement codes or regulations and aim to increase the attractiveness of investing in energy efficiency.

Voluntary instruments, on the other hand, refer to agreements usually between representative bodies of national or regional governments and industry, or even between regional organisations or individual facilities. Such agreements provide a commitment to reduce energy demand by a specified amount over a given period. One example is the UK agreement to phase out incandescent light bulbs, which was superseded by the new EU lighting standards.

Economic instruments aim to correct energy prices either by taxing energy demand and environmental 'bads' such as carbon emissions, or by addressing cost-related challenges (Koeppel and Urge-Vosatz, 2007). Economic instruments can be used to incentivise investment in energy efficiency by offering financial incentives for specific technologies – for example when replacing equipment and for enhanced thermal standards in refurbishment and construction of buildings. Instruments include subsidies or rebates on the purchase of specific products, targeted taxes or tax exemptions and tax credits, financial guarantees and reduced-interest loans (Table 1.2).

The individual instruments outlined above are specific to remedying particular market challenges; no single policy instrument can address multiple market challenges. Given that energy demand in buildings is very complex, fragmented and characterised by multiple challenges, this means that to address energy demand through policy instruments may only be effective when applied in combination with complementary instruments within a package of policy measures (Ott *et al.*, 2005).

It must be remembered that the principal policy view perceives that the success of any instrument or policy package strongly depends on financial incentives, measures that raise awareness and, perhaps most importantly, enforcement (UNEP, 2007). Without enforcement or inducement to conform to the instrument, it is usually argued that it is unlikely to be translated into the desired outcome on the ground.

1.7 Summary and outline of the book

This chapter has taken us on a journey, starting with the fundamentals of why our energy demand is increasing and identifying some of the challenges this demand is causing: climate change, fuel poverty and energy security. Although energy demand is increasing in all sectors, more than one-third of global energy demand (40 per cent in developed countries) is used in buildings for heating, hot water, cooking and electricity to power the ever-increasing number of appliances we seem to need. Moreover, the reasons for increasing demand in buildings are complex, with demographic factors (e.g. increasing populations, smaller households) playing an important role together with our desire for a higher standard of living that has led to the aspiration for larger and warmer homes.

Years of policies and regulations designed to control our energy demand have led to significant progress in improving the efficiencies of our homes and appliances, but this is not proving sufficient to reduce our spiralling energy demand and meet our carbon targets. Conventionally the challenges to reducing energy demand are considered to be due to financial, market and institutional failures. While these obviously play an important role, alternative ways of thinking about these issues are also introduced in this book.

In Chapter 2 we explore ways of reducing the energy demand of buildings during their occupation following the energy hierarchy, which includes both energy conservation and energy-efficiency measures. Although we focus on improving the thermal envelope of buildings, we also examine how escalating electricity demand can be reduced. The demand for heat and electricity does not only occur during the occupation of a building, but throughout its lifecycle, including construction and end-of-life management. Reductions in operational energy demand result in an increase in the relative importance of the energy demand during construction, especially the energy embedded in materials such as insulation, for example. This is explored further in Chapter 3.

Even when buildings are intended to be highly energy efficient, their performance does not always meet expectations. This is particularly the case when the predicted, usually modelled, energy demand is compared with the actual measured demand during its occupation. The underlying reasons for this energy performance gap and potential ways forward to address it are discussed in Chapter 4.

Chapters 2 and 3 mainly focus on new buildings, but if we are to reduce our energy demand we also need to tackle the substantial number of existing buildings. Chapter 5 explores the efficiency of the current building stock and how it can be improved, identifying the challenges that need to be addressed and policies that are currently being used to support retrofitting the homes of fuel-poor households.

Chapter 6 explores a new approach to the design of new (and retrofit of existing) buildings that is gathering a following – Passivhaus, which follows a highly efficient, airtight design. It requires very low levels of heating, relying on internal heat gains from appliances and the occupants, together with solar gain and mechanical ventilation with heat recovery. Improvements in the airtightness of buildings can, however, lead to poor internal air quality and health problems unless adequate ventilation is adopted. Chapter 7 examines these issues and identifies alternative ways forward to address ventilation.

In Chapter 8, we summarise the key findings of this book and discuss what the conventional approaches to reducing energy demand mean in terms of the future of buildings and the future of energy demand. We then explore whether an interdisciplinary systemic approach can help to address some of the challenges.

References

Anderson, K., Bows, A. and Mander, S. (2008) From long-term targets to cumulative emission pathways: Reframing UK climate policy. *Energy Policy*, 36(10), 3714–3722.

Boardman, B., Darby, S., Killip, G., Hinnells, M., Jardine, C.N., Palmer, J. and Sinden, G. (2005) *40% House*, Environmental Change Institute, Oxford.

BP (2014) BP Statistical Review of World Energy Workbook. www.bp.com/content/dam/bp/pdf/Energy-economics/statistical-review-2014/BP-statistical-review-of-world-energy-2014-full-report.pdf, accessed 31 May 2015.

BPIE (2011) Europe's buildings under the microscope: a country-by-country review of the energy performance of buildings. www.bpie.eu/eu_buildings_under_microscope.html#.VWrZcc9VhBc, accessed 31 May 2015.

Brander, M., Sood, A., Wylie, C., Haughton, A. and Lovell, J. (2011) Electricity-specific emission factors for grid electricity. Ecometrica. http://ecometrica.com/assets//Electricity-specific-emission-factors-for-grid-electricity.pdf, accessed 31 May 2015.

Carbon Trust (2005) *The UK Climate Change Programme: Potential Evolution for Business and the Public Sector*, The Carbon Trust, London.

Chalvatzis, K.J. (2012) Approaches to the evaluation of energy security: case studies in the power sectors of Greece and Poland. PhD thesis, University of East Anglia, UK.

Copenhagen Accord (2009) FCCC/CP/2009/L.7. United Nations Climate Change Conference, 18th December, Copenhagen, Denmark.

DECC (2014a) *The Fuel Poverty Statistics Methodology and User Manual*, Department of Energy and Climate Change, London.

DECC (2014b) *UK Energy in Brief Database*, Department of Energy and Climate Change, London.

DECC (2014c) *UK Energy Sector Indicators: Environmental Objectives Dataset*, Department of Energy and Climate Change, London.

DECC (2014d) *Digest of UK Energy Statistics*, Department of Energy and Climate Change, London.

DECC (2014e) *Energy Consumption in the UK: Overall Data Tables, 2014 Update*, Department of Energy and Climate Change, London.

DECC (2014f) *Energy Consumption in the UK: Services Data Tables*, Department of Energy and Climate Change, London.

DECC (2012) *Energy Consumption in the UK*, Department of Energy and Climate Change, London.

DECC (2011) *Renewable Heat Incentive*, Department of Energy and Climate Change, London.

Economist Intelligence Unit (2012) Energy efficiency and energy savings: a view from the building sector. Global Buildings Performance Network. www.economistinsights.com/sites/default/files/down loads/EIU_GBPN_EnergyEfficiency_120921r3.pdf, accessed 31 May 2015.

EEA (2015a) Final energy consumption by sector and fuel assessment. European Environment Agency. www.eea.europa.eu/data-and-maps/indicators/final-energy-consumption-by-sector-8/assessment-2, accessed 5 June 2015.

EEA (2015b) Progress on energy efficiency in Europe. European Environment Agency. www.eea.europa. eu/data-and-maps/indicators/progress-on-energy-efficiency-in-europe-2/assessment, accessed 5 June 2015.

EEA (2012a) Energy efficiency and energy consumption in the household sector. European Environment Agency. www.eea.europa.eu/data-and-maps/indicators/energy-efficiency-and-energy-consumption-5/ assessment, accessed 31 May 2015.

EEA (2012b) Consumption and the environment: 2012 update, the European Environment State and outlook 2010. www.eea.europa.eu/publications/consumption-and-the-environment-2012/at_ download/file, accessed 31 May 2015.

Enerdata (2012) Energy efficiency trends in buildings in the EU: lessons from the ODYSSEE MURE project. www.odyssee-mure.eu/publications/br/Buildings-brochure-2012.pdf, accessed 5 June 2015.

EST (2012) Powering the nation. www.energysavingtrust.org.uk/sites/default/files/reports/Powering thenationreportCO332.pdf, accessed 31 May 2015.

Eurostat (2014a) The EU in the world 2014: a statistical portrait. http://ec.europa.eu/eurostat/ documents/3217494/5786625/KS-EX-14-001-EN.PDF, accessed 31 May 2015.

Eurostat (2014b) Final energy consumption by sector. http://ec.europa.eu/eurostat/statistics-explained/ index.php/Glossary:Primary_energy_production, accessed 5 June 2015.

Eurostat (2013) Glossary. http://ec.europa.eu/eurostat/statisticsexplained/index.php/Glossary: Primary_ energy_production, accessed 5 June 2015.

Flavin, C. and Dunn, S. (1999) A new energy paradigm for the 21st century, *Journal of International Affairs*, 53(1), 169–190.

Friedlingstein, P., Andrew, R.M., Rogelj, J., Peters, G.P., Canadell, J.G., Knutti, R., Luderer, G., Raupach, M.R., Schaeffer, M., Van Vuuren, D.P. and Le Quéré, C. (2014) Persistent growth of CO_2 emissions and implications for reaching climate targets, *Nature Geoscience*, 7, 709–715.

Healy, J.D. (2003) Excess winter mortality in Europe: a cross country analysis identifying key risk factors, *Journal of Epidemiology and Community Health*, 57, 784–789.

IEA (2014) World energy outlook 2014 energy access database. International Energy Agency. www. worldenergyoutlook.org/resources/energydevelopment/energyaccessdatabase/#d.en.8609, accessed 5 June 2015.

IEA (2013) *Technology Roadmap: Energy Efficient Building Envelopes*, International Energy Agency, Paris.

IEA (2012) World energy outlook 2012 factsheet. www.worldenergyoutlook.org/media/weowebsite/ 2012/factsheets.pdf, accessed 31 May 2015.

IEA (2008) *Promoting Energy Efficiency Investments: Case Studies in the Residential Sector*, International Energy Agency, Paris.

IPCC (2014a) *Summary for Policymakers, Climate Change 2014: Impacts, Adaptation, and Vulnerability. Part A: Global and Sectoral Aspects. Contribution of Working Group II to the Fifth Assessment Report of the Intergovernmental Panel on Climate Change*, Cambridge University Press, Cambridge.

IPCC (2014b) Summary for policymakers. In: *Climate Change 2014, Mitigation of Climate Change. Contribution of Working Group III to the Fifth Assessment Report of the Intergovernmental Panel on Climate Change*, Cambridge University Press, Cambridge.

IPCC (2013) *Summary for Policymakers, Climate Change 2013: The Physical Science Basis, Contribution of Working Group I to the Fifth Assessment Report of the Intergovernmental Panel on Climate Change*, Cambridge University Press, Cambridge.

Jiang, L. and Hardee, K. (2011) How do recent population trends matter to climate change? *Population Research Policy Review*, 30, 287–312.

Jiang, L. and O'Neill, B.C. (2009) Household projections for rural and urban areas of major regions of the world, interim report. International Institute for Applied Systems Analysis (IIASA). http://webarchive.iiasa.ac.at/Admin/PUB/Documents/IR-09-026.pdf, accessed 5 June 2015.

Jordan, A., Rayner, T., Schroeder, H., Adger, N., Anderson, K., Bows, A., Le Quéré, C., Joshi, M., Mander, S., Vaughan, N. and Whitmarsh, L. (2013) Going beyond two degrees? The risks and opportunities of alternative options, *Climate Policy*, 13(6), 751–769.

JRC/IE (2009) *Electricity Consumption and Efficiency Trends in the European Union. Status Report*, Joint Research Centre, Institute for Energy, Ispra.

Koeppel, S. and Urge-Vosatz, D. (2007) *Assessment of Policy Instruments for Reducing Greenhouse Gas Emissions from Buildings*, UNEP, Budapest.

Le Quéré, C., Peters, G.P., Andres, R.J., Andrew, R.M., Boden, T.A., Ciais, P., Friedlingstein, P., Houghton, R.A., Marland, G., Moriarty, R., Sitch, S., Tans, P., Arneth, A., Arvanitis, A., Bakker, D.C.E., Bopp, L., Canadell, J.G., Chini, L.P., Doney, S.C., Harper, A., Harris, I., House, J.I., Jain, A.K., Jones, S.D., Kato, E., Keeling, R.F., Klein Goldewijk, K., Körtzinger, A., Koven, C., Lefèvre, N., Maignan, F., Omar, A., Ono, T., Park, G., Pfeil, B., Poulter, B., Raupach, M.R., Regnier, P., Rödenbeck, C., Saito, S., Schwinger, J., Segschneider, J., Stocker, B.D., Takahashi, T., Tilbrook, B., Van Heuven, S., Viovy, N., Wanninkhof, R., Wiltshire, A. and Zaehle, S. (2014) Global carbon budget 2013, *Earth System Science Data*, 6, 235–263.

Liddell, C. and Morris, C. (2010) Fuel poverty and human health: A review of recent evidence, *Energy Policy*, 38, 2987–2997.

Metz, B., Davidson, O.R., Bosch, P.R., Dave, R. and Meyer, L.A. (eds) (2007) *Climate Change: Mitigation of Climate Change: Contribution of Working Group III to the Fourth Assessment Report of the Intergovernmental Panel on Climate Change*, Cambridge, Cambridge University Press.

Nuttall, W.J. (2010) The Euratom Treaty and nuclear energy in the EU-27. In Kondili, E., Lyberopoulos, E. and Kaldellis, J.K., (2013) A Delphi Analysis for the Electricity Supply Security in Greece.

Odyssee (2014) Energy efficiency database. www.indicators.odyssee-mure.eu/energy-efficiency-database.html, accessed 5 June 2015.

OECD/IEA (2014) *Key World Energy Statistics 2014*, International Energy Agency, Paris.

OECD/IEA (2011) *Technology Roadmap: Energy Efficient Buildings: Heating and Cooling Equipment*, International Energy Agency, Paris.

ONS (2014) Excess winter mortality in England and Wales, 2013/14. www.ons.gov.uk/ons/dcp171778_387113.pdf, accessed 31 May 2015.

Ott, W., Jakob, M., Baur, M., Kaufmann, Y., Ott, A. and Binz, A. (2005) Mobilization of energy renewal potential in residential real estate. Research program Energiewirtschaftliche Grundlagen (EWG), Swiss Federal Office of Energy (SFOE), Bern.

Palmer, J. and Cooper, I. (2013a) Great Britain's housing energy fact file. Department of Energy and Climate Change. www.gov.uk/government/uploads/system/uploads/attachment_data/file/48195/3224-great-britains-housing-energy-fact-file-2011.pdf, accessed 31 May 2015.

Palmer, J. and Cooper, I. (2013b) Great Britain's housing energy fact file, tables and graphs. Department of Energy and Climate Change. www.gov.uk/government/statistics/united-kingdom-housing-energy-fact-file-2013 accessed 5 June 2015.

Peters, G.P., Andrew, R.M., Boden, T., Canadell, J.G., Ciais, P., Le Quéré, C., Marland, G., Raupach, M.R., and Wilson, C. (2012) The challenge to keep global warming below 2°C, *Nature Climate Change*, 3, 4–6.

Public Health England (2013) Cold weather plan for England: making the case – why long-term strategic planning for cold weather is essential to health and wellbeing. www.gov.uk/government/

uploads/system/uploads/attachment_data/file/365269/CWP_Making_the_Case_2014_FINAL.pdf., accessed 31 May 2015.

Raupach, M.R., Gloor, M., Sarmiento, J.L., Canadell, J.G., Frölicher, T.L., Gasser, T., Houghton, R.A., Le Quéré, C. and Trudinger, C.M. (2014) The declining uptake rate of atmospheric CO, *Biogeosciences*, 11, 3453–3475.

Royal Academy of Engineers (2010) Engineering a low carbon built environment: the discipline of Building Engineering Physics, http://www.raeng.org.uk/publications/reports/engineering-a-low-carbon-built-environment.

SBCI (2007) *Buildings and Climate Change: Status, Challenges, and Opportunities*, United Nations Environment Programme, Sustainable Buildings and Construction Initiative, Paris.

Shorrock, L. and Utley, J. (2003) *Domestic Energy Fact File*, Building Research Establishment, Watford.

Sovacool, B.K. and Mukherjee, I. (2011) Conceptualizing and measuring energy security: A synthesized approach, *Energy*, 36(8), 5343–5355.

Thomson, H. and Snell, C. (2013a) Quantifying the prevalence of fuel poverty across the European Union, *Energy Policy*, 52, 563–572.

Thomson, H. and Snell, C. (2013b) Energy poverty in the EU policy brief. http://fuelpoverty.eu/wp-content/uploads/2013/06/Final-energy-poverty-policy-brief-EU-Energy-Week-13.pdf accessed 5 June 2015.

UN (2014) Population facts: population ageing and sustainable development. United Nations Department of Economic and Social Affairs, Population Division. www.un.org/en/development/desa/population/publications/pdf/popfacts/PopFacts_2014-4.pdf, accessed 5 June 2015.

UN (2013) World population ageing. United Nations Department of Economic and Social Affairs, Population Division. www.un.org/en/development/desa/population/publications/pdf/ageing/World PopulationAgeing2013.pdf, accessed 5 June 2015.

UN (2012) Population database and projections. United Nations Department of Economic and Social Affairs, Population Division, Population Estimates and Projections Section. http://esa.un.org/unpd/wpp/unpp/panel_population.htm, accessed 5 June 2015

UN Millennium Project (2005) *Investing in Development: A Practical Plan to Achieve the Millennium Development Goals*, UNDP, Paris.

UNEP (2007) *Buildings and Climate Change: Status, Challenges and Opportunities*, UNEP, Paris.

UNFCCC (2011) Establishment of an ad hoc working group on the Durban Platform for Enhanced Action. https://unfccc.int/files/meetings/durban_nov_2011/decisions/application/pdf/cop17_durbanplat form.pdf, accessed 31 May 2015.

Ürge-Vorsatz, D., Mez, L., Miladinova, G., Antipas, A., Bursik, M., Baniak, A., Jánossy, J., Nezamoutinova, D., Beranek, J. and Drucker, G. (2003) *The Impact of Structural Changes in the Energy Sector of CEE Countries on the Creation of a Sustainable Energy Path*, European Parliament, Luxembourg.

Vreuls, H. (2005) *Evaluating Energy Efficiency Policy Measures and DSM Programmes: Volume I: Evaluation Guidebook*, International Energy Agency, Paris.

WHO (2014) *WHO Guidelines for Indoor Air Quality: Household Fuel Combustion*, WHO, Geneva.

WHO (2007) *Housing, Energy and Thermal Comfort: A Review of 10 Countries Within the WHO European Region*, WHO, Geneva.

Wilkinson, P., Landon, M., Armstrong, B., Stevenson, S., Pattenden, S., McKee, M. and Fletcher, T. (2011) *Cold Comfort: The Social and Environmental Determinants of Excess Winter Deaths in England, 1986–1967*, The Policy Press and the Joseph Rowntree Foundation, Bristol.

Wilkinson, P., Smith, K.R., Davies, M., Adair, H., Armstrong, B.G., Barrett, M., Bruce, N., Haines, A., Hamilton, I., Oreszczyn, T., Ridley, I., Tonne, C. and Chalabi, Z. (2009) Public health benefits of strategies to reduce greenhouse-gas emissions: household energy, *Lancet*, 374, 1917–1929.

Williams, J. (2007) Innovative solutions for averting a potential resource crisis: the case of one-person households in England and Wales, *Environment, Development and Sustainability*, 9, 325–354.

World Bank Group (2005) *Energy Security Issues*, World Bank, Washington, DC.

York, R. (2007) Demographic trends and energy consumption in European Union nations, 1960–2025, *Social Science Research*, 36, 855–872.

2 Reducing energy demand

2.1 Introduction

It is vital that the energy demand of buildings is reduced, especially if, as discussed in Chapter 1, we are going to meet our carbon emissions targets. It has been estimated that the energy demand of both existing and new buildings can be reduced by 70–80 per cent, with the potential for reducing the demand of new buildings being greater than that of existing buildings (Huovila, 2007). Most of the technologies and design strategies required to instigate this reduction already exist and are considered to be 'mature' (Metz *et al.*, 2007; Nicol and Roaf, 2007). By 'mature' we mean the technology or design strategy is one that has been in use for a period of time, though not necessarily in widespread use, and is now relatively stable and well understood.

This chapter discusses the two main approaches to reducing the energy demand of buildings: energy conservation and energy efficiency. It also introduces the principal concepts required to understand how heat moves through a building's fabric, as well as briefly discussing the necessary steps for decarbonising a building's energy supply. The reduction of energy is nearly always considered from an engineering or technical perspective in which technology is used to improve the energy efficiency of supplying the required building services (e.g. heating, cooling and lighting). Although technological improvements are important, this technical approach can exclude broader influences, such as how society is organised, which also shape energy demand. For instance, the deployment and effectiveness of technological improvements are influenced by institutional arrangements and political decisions, in addition to how the technologies are actually operated. Although trying to change behaviour is included in conventional energy conservation approaches, this is usually undertaken in terms of trying to persuade people to adopt certain ways of doing things by providing them, for example, with information. This chapter will explore behaviour changes that interface with the use of technology such as energy display monitors. The broader institutional and social aspects of reducing energy demand are introduced in more detail in Chapters 4, 5 and 6.

2.2 The energy hierarchy

To achieve low energy demand, and to minimise carbon emissions, a basic guiding framework is required. Based on the waste hierarchy (DEFRA, 2011), which in its simplest form is to reduce, reuse and recycle, the energy hierarchy has been developed to provide a clear, practical framework that has been widely adopted (e.g. Greater London Authorities Energy Plan (GLA 2014)). There is, however, no standard definition of the energy hierarchy, and many interpretations exist. In this book the energy hierarchy is defined as having a sequence of five priorities based on the principle of reducing energy demand first before addressing issues of decarbonisation and the availability of the energy supply (Figure 2.1)

The first two priorities of the energy hierarchy are concerned with managing (i.e. reducing) the demand for energy. The first priority, energy conservation, is the reduction or elimination of unnecessary energy demand, or eliminating waste (see Section 2.3).

The second priority is energy efficiency. There are two main aspects to energy efficiency: the efficient conversion of energy (such as the generation of electricity in power stations) and the efficient use of energy. In the built environment context we are concerned with the efficient use of energy, which is defined here as technological improvements that reduce the amount of energy required to provide the same, or a better, level of service (see Section 2.4).

Energy conservation and energy efficiency are often confused, but they are not the same. Energy conservation involves using less or going without a service to save energy, whereas energy efficiency is concerned with using less energy without reducing the level of service. For example, turning off a light is energy conservation, while replacing an incandescent light bulb with a bulb that uses less energy to produce the same amount of light (i.e.

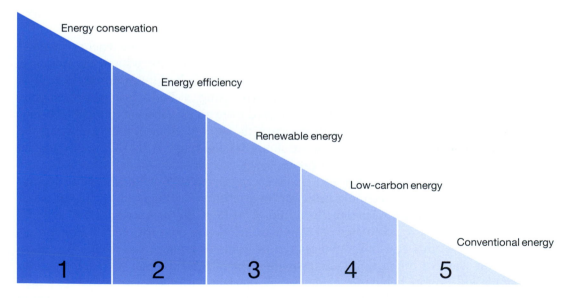

Figure 2.1 A graphical representation of the energy hierarchy, showing relative importance of priorities from left to right.

a compact fluorescent lamp (CFL) or a light emitting diode (LED) lamp) is energy efficiency. Energy conservation and energy efficiency are linked but neither automatically guarantees energy reduction, due to the rebound effect (Section 4.8.3).

The remaining priorities are concerned with energy supply. The third priority refers to the substitution of conventional energy with renewable energy resources such as wind, solar thermal, photovoltaics and biomass (Table 2.1). These are described in many

Table 2.1 Summary of technical strategies for reducing energy and carbon emissions in new homes

Strategy	Energy hierarchy				Type of energy	
	Demand side		Supply side		Heat	Power
	Energy conservation	Energy efficiency	Renewable energy	Low carbon energy		
Behaviour:						
Switching off equipment when not in use	X				X	X
Information provision	X				X	X
Promotional activities	X				X	X
Fabric:						
Insulation		X			X	
Ventilation		X			X	
Airtightness		X			X	
Design:						
Passive solar design	X	X			X	
Thermal mass	X	X			X	
Technology: renewable						
Photovoltaic			X		X	X
Solar thermal			X			
Biomass			X		X	X
Biofuels			X			X
Wind			X			X
Hydro			X			X
Technology: conventional fuels						
Heat pump			X	X	X	
Micro-CHP	X		X	X		
Co-heating/district heating			X	X	X	X
Post-construction:						
Commissioning	X				X	X
Control	X			X	X	X
Meters and monitors	X				X	X

Source: Monahan (2013).

good-quality texts on energy supply (e.g. Boyle, 2012; Elliott, 2013), so only small-scale micro-generation, produced to reduce the energy demand of individual buildings, is discussed briefly in this book (Section 2.5). The final two priorities in the energy hierarchy involve conventional low-carbon energy systems and fuels, plus improving the efficiency of energy production and fuel switching. These include low-carbon energy sources such as nuclear power or more efficient energy generation such as combined heat and power (CHP). Relating to energy supply rather than demand, these fall outside the scope of this book and are not discussed further.

There are a range of strategies for reducing energy and carbon emissions from buildings (Table 2.1). Many of these strategies and associated concepts apply equally to all buildings, new and existing. However, existing buildings have a range of issues and problems intrinsic to them, which are discussed in Chapter 5.

The energy hierarchy prioritises energy demand through minimising waste and improving efficiency before engagement with supply-side activities. As such, it offers an effective framework to guide energy management and decision making in reducing energy in the built environment.

2.3 Energy conservation in buildings

Energy conservation typically involves individuals performing actions that save energy – for example, switching off lights or appliances when not required, or turning down the thermostat for a heating system. Energy conservation is a common target of mainstream behaviour change programmes (e.g. social marketing campaigns that show saving energy to be 'cool'). Other types of initiatives focus on convincing individuals to conserve energy more indirectly, perhaps trying to empower them by providing information on how much energy they use or which types of equipment in their building require the most energy. This information can be provided by energy display meters, which are discussed later in this section. Another approach is to provide building occupants with new technologies to enable them to perform a particular energy-saving behaviour that they were not doing previously, even if they had wanted to. An example of this is improved heating controls that enable occupants to more effectively match heating system operation to occupancy patterns. It is this philosophy that often provides a rationale for a move to 'smart' homes, a topic we explore later. In addition, we reflect upon how frustrations with 'convincing' occupants to perform energy conservation behaviours have led to the advocacy of (smart) automation, where the burden of energy conservation is passed from the occupant to the technology and its designer.

2.3.1 Energy conservation behaviour

Changing the behaviour of energy end-users has long been recognised as a low-cost energy conservation method that has potential to produce significant energy savings both in the home and at work, and has therefore been the target for activities such as awareness raising campaigns and incentive schemes. Targeted behaviours typically include switching devices and appliances off when they are not being used, setting timings of equipment to meet the patterns of the day and using controls to optimise settings.

A significant proportion of energy demand is unnecessary or 'wasted' energy. For example, a recent study by the EST (2012) found that standby electricity demand amounted to 9–16 per cent of total UK residential electricity demand. Standby power refers to the power consumed by an appliance or device while it is switched off but still drawing power (i.e. in standby mode). Standby power can be eradicated by switching off the specific appliance at the plug or disconnecting from the power point. Such simple actions to conserve energy can radically reduce demand. Table 2.2 illustrates this by providing a selection of different energy conservation behaviours and the estimated energy savings each action can produce for a UK household if widely adopted. For example, switching appliances off when not in use is estimated to save an average 343 kWh per year for a household (EST 2012), 4,349 GWh per year nationally if adopted by 50 per cent of UK households. Numerous other examples that illustrate the potential impact of energy conservation are detailed by Palmer *et al.* (2012). The report estimates the potential energy savings that could be achieved if all UK households adopted simple changes to their everyday behaviour; turning off the TV would, for example, save 101,325 GWh/year.

Table 2.2 Energy conservation behaviours and estimated savings potential for households (kWh/household/year; GWh/year) and businesses in the UK

Behaviour action	Per household (kWh/household/year)	UK national (GWh/year)
Household[a]		
Switch appliances off to avoid standby[b]	343	4,349
Turn off TVs	49	759
Turn down thermostat by 1°C (from 19 to 18°C)	1,528	16,484
Delay start of heating from October to November	667	10,572
Use dishwasher on eco setting	182	1,125
Only use dishwasher when full	104	148
Avoid using tumble dryer, air dry instead	364	1,673
Wash clothes at 40°C or lower	70	256
Maintain fridges and freezers (regular defrost and clean coils)	105	1,335
Only fill kettle with required amount	83	1,200
Turn off lights when not in use	134	2,128
Cook with lids on saucepans	124	588
Commercial:[c]		
Switch off non-essential lighting out of business hours	10 per cent of lighting costs	
Alter timing of HVAC to meeting comfort demands of the working day	20 per cent of heating and cooling costs	
Set thermostats correctly	8 per cent of heating costs per 1°C reduction	
Turn off unnecessary equipment when not in use	5 per cent of energy costs	

Source: (a) Palmer *et al.* (2012); (b) EST (2012); (c) Williams and Tipper (2013).

Changing energy use behaviour also holds true for the commercial sector. For example, according to the UK's Carbon Trust (Williams and Tipper 2013), encouraging employees to adopt low-energy behaviours is estimated to save £300 million and over six million tonnes of carbon. The report recommends an investment of 1–2 per cent of energy spend on raising employee awareness and promoting energy conservation activities, which it argues could lead to a 10 per cent reduction in energy demand (Williams and Tipper, 2013). For example, switching off non-essential lighting outside of business hours is estimated to reduce lighting energy costs by 10 per cent (Williams and Tipper 2013) (Table 2.2).

Energy conservation initiatives conventionally rely on communication and engagement by providing information and promotional techniques in campaigns that aim to raise awareness and institute energy conservation behaviours. Typically these campaigns rely on media, such as posters or advertisements to raise awareness and promote specific behaviours. The Student Switch-Off campaign (Section 2.7.1, case study 1) is an example of a UK-based energy-efficiency awareness raising campaign.

Despite significant efforts to introduce energy conservation behaviours, and a large and growing literature investigating the effectiveness of these techniques, the evidence suggests that the majority of campaigns are never as effective as hoped for, or may not have any impact at all (Abrahamse *et al.*, 2005). More pertinent still, such campaigns may not institute a 'permanent' change in behaviour, with well-documented energy 'creep' after the initial gains have been made (Abrahamse *et al.*, 2005). This occurs where the initial gains made are eroded by increasing energy demand over time till energy demand is at the pre-campaign level or higher.

2.3.2 *Visual display meters*

We will take a moment to discuss one particular type of information provision (and 'awareness' raising) that has been given considerable attention and support in recent years: the visual display meter (sometimes also referred to as an 'in home display'). A visual display meter provides a real-time display of the energy (usually electricity, but could be gas) being used in a building, providing the occupants with, usually in graphic form, information about their current energy use at that moment in time. Additional information on the carbon emissions and costs may also be given, depending on the particular product used. This is especially pertinent in households where metering equipment is hidden away or located outside the home. There is a growing body of literature that has investigated the impact of providing such real-time energy use information to occupiers. The picture is unclear. On the one hand, visible metering, monitoring and providing feedback to energy users has been shown to reduce energy demand in the home, with the provision of energy display meters being predicted to reduce energy demand by 5–15 per cent (Darby 2006). Visible display metering may also have a role in optimising the use of on-site renewable energy generation. For example, Keirstead (2007) found that the presence of visible monitoring of outputs from PV systems facilitated a shift in energy demand which was attributed to a change in energy awareness.

On the other hand, there is a more recent body of literature that shows the reduction in energy demand may be overstated, with the savings being much lower (CPA, 2014). For example, a large-scale UK study indicated the savings could be only 3 per cent (Hargreaves *et al.*, 2013), while there is also evidence to suggest that the savings may not be sustained in the long term (van Dam *et al.*, 2010). A study on real-time energy display monitors found the energy savings from the use of these devices dropped off quickly after an initial 'honeymoon' period in which the device was a novelty (Hargreaves *et al.*, 2013).

Visual display meters are sometimes used in combination with smart meters, discussed below.

2.3.3 Controls and smart homes

One way of avoiding the difficulties associated with changing people's behaviour is to increase the automation of energy systems, creating a 'smart building'. A first step towards this is for the controls on heating, cooling and ventilation systems to be more straightforward to use so as to enable occupants to control the energy they use easily, and hopefully reduce their demand. But relying on people to use the controls in this way can be difficult, so an alternative approach is for them to be automated. Rather than relying on the active participation of users to adjust their level of ventilation or to turn heating on and off, controls are used to specify when and for how long a system is operated for or to what level of service, and thus, it is hoped, optimise the use and energy demand. Automated controls can be relatively simple, such as daylight sensors or movement sensors for lighting. Alternatively, they can be more complex – for example, a domestic heating system's thermostatic controls may include functions that allow the boiler to control the radiators differently in different rooms or areas, to multiple timed periods for each day of a week and to different temperatures.

Control systems obviously need to be used effectively to ensure the system they are controlling achieves the desired energy conservation aims. There is a growing body of evidence, however, that suggests this is not occurring. For example, occupants may not know or understand the functions of the different elements of the control systems or how they can be operated (Liao *et al.*, 2005). Combe *et al.* (2010) found 66 per cent of occupiers living in a low-carbon housing development were unable to programme their heating systems controls; their results indicated that poor design is a key issue. Complexity, inconveniently placed controls and badly designed interfaces and menus on controls have also been cited as reasons why occupants fail to use controls effectively. This results in users either entering unsuitable settings or avoiding interacting with the controls entirely, resorting to either switching the system on and off when required or putting up with inconvenience or discomfort (Boait and Rylatt, 2010). This suggests that the design of controls is at fault rather than the capabilities of the user. Improving the usability of controls through a more user-centred design of control products has been suggested to be key to addressing this problem (Stevenson *et al.*, 2013).

A smart home goes further than controls on individual systems, involving an integrated system that contains a range of IT equipment linked to devices that will automate and manage the control of energy systems (e.g. heating, ventilation and lighting) and appliances (e.g. washing machines) in our homes and offices. A smart home is sometimes visualised as meeting the future needs of an ageing society in that it will contribute to 'better' and 'assisted' living (Rashidi and Cook, 2009). However, the smart home is mainly portrayed as a way of reducing our energy demand (Wilson *et al.*, 2015), particularly as part of a carbon-constrained future. It should be noted that although the term 'energy' is used here, the smart home concept is predominantly concerned with electricity. Similarly, although much of the literature refers to smart 'homes' it often equally applies to other buildings.

Much research associated with the future of energy mentions the innovations of the smart grid. Although this might appear to be linked to energy supply rather than demand, the end 'node' of the smart grid has been identified as the 'smart home' (Darby, 2010). We will, therefore, briefly explore the concept of the smart grid.

The conventional electricity grid supports a predominantly one-way flow of electricity from a few large-scale power stations to numerous, generally small-scale users. But in many countries this conventional (and often ageing) infrastructure no longer meets our needs. There are now many more small-scale generators on the distribution network, many of which are intermittent (i.e. the output varies with the availability of resources such as the sun and wind). In addition there is the need for the management of demand, in which, in response to peaks in electricity demand, electricity is reduced to non-essential users (e.g. refrigeration units) for short periods of time. Improved digital technology that allows for two-way communication between the electricity suppliers and their customers is needed. This smart grid will operate in a similar way to the internet, with a network of controls, computers, automation and new technologies and equipment working together, but with the aim of responding rapidly to changes in supply and demand. The sensing along the transmission lines is what makes the grid 'smart'.

An essential part of a 'smart home' is a 'smart meter', which may or may not include a visual display. A smart meter is an interface between the householder and their electricity supplier, which allows the transfer of information to and from the electricity supplier with real-time information on supply and demand across the distribution network (Palansky and Kupzog, 2013). In the future, an energy supplier, usually in return for a reduced tariff, will be able to use this interface to reduce its customers' peak electricity demand, by temporarily cutting the supply to non-essential appliances, such as a fridge, for a short time. In a similar way, plug-in electric cars could be used for temporary electricity storage or supply.

Increasingly, sophisticated building management systems may remove the reliance on changing individual behaviour of building occupants in energy conservation, but they are in turn reliant on human behaviour, as there is a need to change the behaviour of other 'actors', such as the people that influence the design, purchase and installation of technologies such as controls or smart meters.

Energy conservation often appears to be a quick, easy and cheap method of reducing energy demand in buildings. Studies have shown that this is not actually the case in reality, as approaches that rely on changing the behaviour of occupants routinely fail to achieve their desired outcome. Technical approaches may be easier. Instead of trying to change the behaviour of a population, might it be easier to devolve this responsibility by increasing the automation of building management systems and by introducing energy-efficiency measures?

2.4 Energy efficiency in buildings

Energy efficiency is a 'measure of energy used for delivering a given service. Improving energy efficiency means getting more from the energy that we use' (DECC, 2012, p. 6). For example, insulating a building or replacing a cathode ray tube (CRT) television with a liquid crystal display (LCD) television, or replacing single-glazed with double-glazed windows. As this suggests, there are two main threads to energy efficiency in buildings. The first relates to the building itself and its thermal efficiency, while the second to the technologies that already exist, or that occupants bring into the buildings, e.g. lights and appliances.

Here we focus on the building's thermal efficiency, introducing what is known as the fabric first approach. This is followed by an introduction to reducing electricity demand in buildings for lighting and appliances.

2.4.1 The fabric first approach

Energy efficiency in buildings is typically achieved by following a 'fabric first' approach. This approach focuses on minimising the rate of heat loss from a building through its thermal envelope. Reducing heat loss is achieved through using insulation, thermally efficient windows and doors, and minimising the unwanted air that leaks through gaps in the fabric (termed infiltration). By using these techniques, the heat loss from a building can be reduced to a point where the energy demand for space 'conditioning' (i.e. heating or cooling) becomes almost negligible (Roaf *et al.*, 2007). These are the core design principles of the Passivhaus approach (Chapter 6).

This section describes heat loss in more detail and introduces the four key elements that contribute to heat loss: insulation, thermal bridging, airtightness and ventilation. This includes consideration of building design and the specification of materials. For example, in passive solar design, the building elements (walls, floors and windows) are used to collect, store and distribute solar energy. This is termed thermal mass, the ability of a material to absorb heat. High-density materials, such as concrete, clay, stone and water, require a large amount of heat to change temperature and are therefore said to have a high thermal mass. In contrast, lightweight materials such as timber are said to have low thermal mass. High thermal mass materials can be used to attenuate fluctuations in internal temperatures.

2.4.2 An introduction to heat loss

Heat will always flow down the gradient from hot to cold areas. Consequently, all buildings lose or gain heat from the surrounding environment. The most effective way of reducing the amount of energy required to keep a building thermally comfortable is to reduce the amount of heat loss rather than achieving the desired level of thermal comfort through increasing the amount of heating required to compensate for lost heat.

Heat loss is defined here as the amount of heat transmitted from the inside of a building to the outside environment through its thermal envelope – the units are watts per square metre (W/m^2). A building's thermal envelope includes walls, roofs, floors and windows (Figure 2.2). In calculating heat loss it is only these elements which make up the building's thermal envelope that are of interest.

Heat is lost from a building in two ways: through the fabric and through ventilation. Fabric heat losses are those that occur by conduction through the building's fabric (i.e. walls, roofs, floors and windows) and via convection and radiation from the fabric's

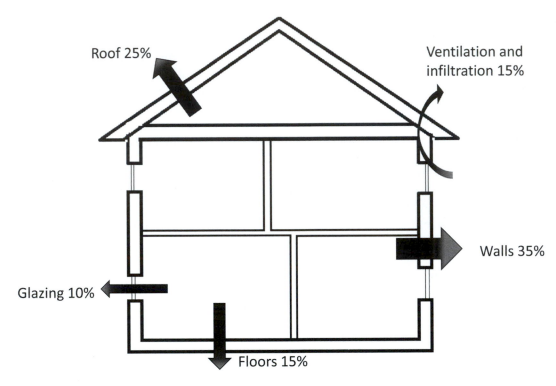

Figure 2.2 Illustration of heat loss through built fabric and ventilation (%).
Source: EST (2015).

surfaces (Beggs, 2010). In contrast, ventilation heat losses are those that occur by convection when warm air is lost and is replaced by cold air that requires heating.

Unsurprisingly, most buildings are not airtight. Indeed, a totally airtight building with no air changes is unachievable and undesirable. In addition to ventilation that is necessarily designed into a building (including opening windows or mechanical ventilation systems), there is also unwanted air movement through gaps, cracks and holes in the building's fabric (termed leakage). Typically arising from flaws in the construction design (e.g. thermal bridging) or poor workmanship, these defects result in undesirable 'air leakage'.

Both fabric and ventilation heat loss rates are directly proportional to the difference in temperature between the internal space and the outside environment. The greater the temperature difference, the greater the flow of heat. The easiest way to reduce heat loss and improve the thermal performance of a building is to insulate it and reduce the rates of air leakage. This requires a holistic approach that takes into consideration not just levels of insulation, but also how the building is constructed.

2.4.3 Insulation

Conventionally, in striving for energy efficiency, building designers have concentrated on improving the thermal performance of individual building elements (i.e. exterior walls, floors and roofs) by improving levels of thermal insulation. The aim of insulation is to improve the resistance of heat transfer across the thermal envelope. Some materials are better at resisting the flow of heat than others and these properties are exploited to provide insulation (see Table 2.3 for details of key technical terms and Table 2.4 for common insulation materials). In cold climates the aim of insulation is to reduce heat flow out of the building. Conversely, in hot climates the main aim of insulation is to minimise the flow of heat into the building.

So, how much insulation is enough insulation? This depends on local regulatory requirements, local climate, energy costs, project budget and personal preference. Many countries have building regulations that specify minimum requirements and recommend insulating above these minimum requirements. For example, the UK's *Building Regulations Part L: The Conservation of Fuel and Power* has a maximum allowable U-value for walls of 0.20 $W/m^2 \cdot K$. (DCLG, 2013).

Table 2.3 Technical terms relating to thermal insulation including thermal conductivity, thermal resistance and thermal transmittance

Technical term	Units	Illustration	Description of key points
Thermal conductivity lambda (λ) or k-value	$W/m \cdot K$		Standardised measure of how easily heat flows through a material • independent of material thickness • the lower the k-value the better the thermal performance • e.g. thermal conductivity of polyurethane insulation (PUR) is 0.02 $W/m \cdot K$ compared with brick, which is 0.6–1.0 $W/m \cdot K$ (Table 2.4)
Thermal resistance R-value	$m^2 \cdot K/W$		Measure of how much heat loss is reduced through a given thickness of a material • calculated by $R = l/\lambda$ where l is the thickness of material and λ is the thermal conductivity in $W/m \cdot K$ • the thermal resistance for a series of layers in an element is calculated by summing together the individual R-values • the higher the R-value the better the thermal performance • can be used to determine the effectiveness of a material as insulation
Thermal transmittance U-value	$W/m^2 \cdot K$		A measure of the overall rate of heat transfer through a square metre of a material or a construction element • calculated by taking the reciprocal of the R-value (1/R-value) plus conduction, convection and radiation at the surface and thermal bridging • the lower the U-value the better the thermal performance

Table 2.4 Thickness (mm) required to achieve U-value of 0.25 (W/m²·K) and thermal conductivity (W/m·K) of common construction materials

Material category	Type	Thickness (mm) required to achieve U-value of 0.25 (W/m²·K)	Thermal conductivity (W/m·K)
Advanced materials	Aerogel		0.008
	Vacuum insulated panels		0.0013–0.0014
Polyurethane (PUR)	Foil faced with pentane (up to 32 kg/m³)	75	0.020
Polyisocyanurate (PIR)	Foil faced (up to 32 kg/m³)	80–85	0.022–0.023
Phenolic foam (PF)	Foil faced (up to 32 kg/m³)	75–85	0.02–0.023
Expanded polystyrene (EPS)	Up to 30 kg/m³	115–165	0.03–0.045
Extruded polystyrene (XPS)	Extruded with CO_2	95–140	0.027–0.037
Wool and fibre	Glass wool	135–180	0.03–0.044
	Stone wool	150–170	0.034–0.038
	Sheep's wool	150–215	0.034–0.054
	Cellulose fibre	150–190	0.035–0.046
	Hemp fibre	165	0.039
	Polyester fibre	150–180	0.035–0.044
	Wood fibre	145–225	0.039–0.061
Other	Hemp lime monolithic	260	0.067
	Cotton	165–170	0.049–0.040
	Cork	155–200	0.041–0.055
	Vermiculite	205	0.039–0.060
	Perlite	150	0.051
	Cellular glass	140–185	0.038–0.050
	Thermal wall linings	N/A	0.04–0.063
	Strawboard	295	0.081
	Straw bale monolithic	175–235	0.047–0.063
Brick common		N/A	0.60–1.00
Wood		N/A	0.12–0.14
Breeze block		N/A	0.10–0.20
Concrete		N/A	0.10–1.80
Steel		N/A	50.00

Source: CIBSE (2006).

In some countries (e.g. Russia) energy costs are very low, disincentivising reductions in heat loss. However, in other locations (e.g. Scandinavia) energy may be expensive and the climate cold, thereby incentivising thermally efficient building with high levels of insulation.

Increasing insulation has, of course, cost implications; materials with a greater thermal performance for a given thickness are generally more expensive. For example (at current UK retail prices, 2015), to achieve a U-value of 0.25 $W/m^2 \cdot K$ would require approximately 180 mm thickness of mineral wool insulation, costing approximately £8–10 per square metre, or approximately 80 mm thickness of rigid polyurethane (PUR) insulation costing £17–19 per square metre, 48 per cent more.

Increasing the insulation specification may also have implications for the sizing of other items in the construction, such as the widths of structural elements like timbers. Furthermore, the widths of walls may need to be increased to accommodate a greater depth of insulation. This will result in the wall taking up a larger area which will have to be compensated for by either increasing the footprint of the building or by reducing the internal area of the building.

The personal preferences of the person responsible for commissioning the building may also have implications for how much insulation is 'enough'. For example, the commissioner of the building may be environmentally motivated or want a building that is cheap to heat and be prepared to offset these long-term savings by investing upfront in the increased capital cost of construction. Conversely, the commissioner may have no interest in the energy demand of the building and be motivated by up-front profit and want a building that is cheap to construct. This is known as a split incentive and is discussed further in Chapter 4. In this situation the minimum insulation required by regulations (if regulations exist) will be 'enough'.

2.4.4 Thermal bridging

As heat flows down the thermal gradient it will flow along the path of least resistance from the heated space to the outside (and vice versa in warm climates). This movement along the path of least resistance includes moving through any unintended gaps, flaws in the structure or places where two layers with different thermal conductivities meet. This is termed 'thermal bridging' (it may also be referred to as a cold bridge) and occurs, for example, at the junctions between walls, floors, roofs, windows, doors and balconies (Figure 2.3). Thermal bridges can contribute significantly to poor indoor environments, resulting in significant heat loss, reduced interior surface temperatures, condensation and moisture problems within the construction, resulting in the growth of mould (Ward 2006).

An effective solution to heat loss is to wrap the outside of the building with insulation. While this may be relatively easy when designing new buildings, solving thermal bridging in existing buildings is not easy. However, a world in which all buildings are uniform, regular boxes would be very dull indeed; there will always be windows, doors, balconies and other architecturally necessary or interesting design items that may cause thermal bridging in new buildings. These can be avoided at the design stage by installing a thermal break, which is a section of material with a low thermal conductivity placed between the two materials to slow the transfer of heat.

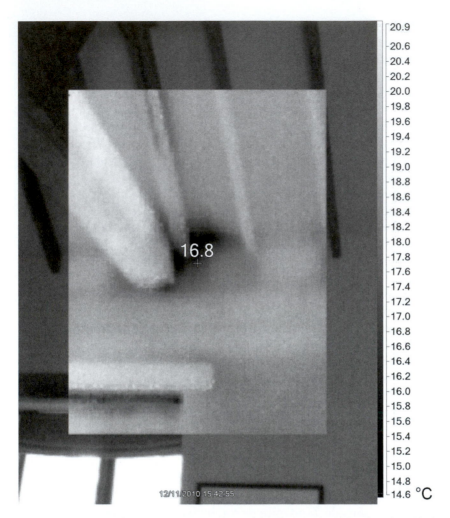

Figure 2.3 Thermal image showing thermal bridging where roof and wall elements meet. Scale (°C) showing warmest areas coloured white, coldest areas coloured black.

Image: J. Monahan.

2.4.5 Airtightness

As levels of insulation increase in buildings, heat losses related to air leakage become more significant. As a consequence, the higher the energy efficiency of a building the more important airtightness becomes. Furthermore, unwanted air leakage through the structure means the expected thermal performance of a building will not be met, resulting in an energy performance gap (Chapter 4).

An airtightness test provides a crucial measure of the airtightness of a building, indicating the location and extent of uncontrolled air paths. In an airtightness test all vents, windows and doors are closed and a fan unit is used to pressurise and depressurise the building. This allows the measurement of the amount of air that leaks into and out of the building and therefore quantifies the amount of draughts that may exist. The location of the leakages can be identified by the use of a smoke 'pencil', with the smoke showing the air movement. The results of an airtightness test can be expressed in two ways:

1   *Air permeability* – rates are measured in cubic metres of air leaking through a metre square of external fabric per hour at a pressure difference of 50 Pa (with units $m^3/hr/m^2$ of total surface area).
2   *Air changes per hour (ach)* – measured as the number of air changes occurring per hour from a building of a given volume.

Airtightness is strongly influenced by construction type and design. For example, studies have shown that concrete buildings tend to be more airtight than brick masonry, which in turn are more airtight than timber frame dwellings (Johnston *et al.*, 2004). In the UK a shift in construction practices, from a wet-plastered internal wall finish to plasterboard dry lining during the 1980s, significantly reduced the airtightness of new-build homes (Figure 2.4).

So what level of airtightness is right? This is largely dependent upon local mandatory and voluntary codes, local climate and the preferences of a building's commissioner or developer. Airtightness is, therefore, prioritised as an energy-efficiency solution differently depending upon the context. Airtightness standards vary widely, with countries such as Sweden and Canada, which experience cold winters, tending to have the most stringent requirements (Table 2.5). In the UK the current regulatory requirement for new housing is a maximum air leakage rate of 10 $m^3/m^2/h$ at 50 Pa (Table 2.5). A recent UK study that examined the results from testing the airtightness of 1,293 new dwellings found the average to be better than this at 6 $m^3/m^2/h$ at 50 Pa (Phillips *et al.*, 2011). However, to put this into context, Sweden has a minimum standard of 1.5–3.0 $m^3/m^2/h$ at 50 Pa. Voluntary standards tend to require higher levels of airtightness than mandatory standards. For example,

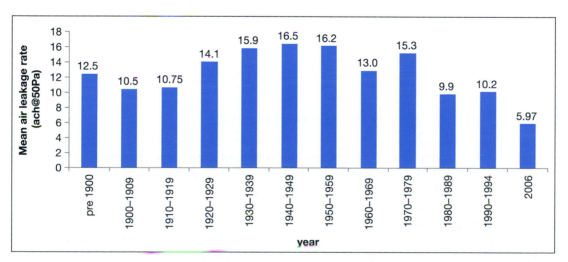

Figure 2.4 Dwelling age (years) and airtightness (ach at 50 Pa) of UK housing (pre-1900–2006). Source: Lowe *et al.* (1997), Stephen (2000), Pan (2010).

Table 2.5 Examples of maximum airtightness standards from different countries (ach at 50 Pa)

Country	Air changes per hour (ach at 50 Pa)
Netherlands	6
Switzerland	3.6
Germany	1.8–3.6
Denmark	2.8
Canada (Super-E)	<2
USA	3–5

Source: Pan (2010).

Passivhaus requires an air change rate of less than 0.6 ach at 50 Pa (for further details see Chapter 6).

As airtightness progressively improves, the provision of adequate ventilation becomes increasingly critical. Indeed, an oft-quoted adage is *'Build tight, ventilate right'*. Historically, buildings have relied on windows and infiltration (experienced as draughts) to provide adequate ventilation. Increasing levels of airtightness and the radical reduction of infiltration as a source of fresh air means ensuring adequate ventilation has become essential in low-energy buildings. Good levels of fresh air are essential to ensuring a healthy indoor environment. The health effects of poor ventilation in buildings are well documented, and the term 'sick building syndrome' has been coined to refer to a range of symptoms, including respiratory complaints, irritation, fatigue, respiratory diseases and cancer (Crump *et al.*, 2009). As buildings become increasingly airtight it becomes more difficult to rely on traditional passive ventilation approaches, such as windows and vents, to provide cross ventilation or passive stack ventilation. As a consequence, ventilation systems have become increasingly mechanised. In particular, mechanical ventilation with heat recovery (MVHR) is increasingly being used in new buildings in order to satisfy the goals of minimising energy demand without compromising indoor air quality (Crump *et al.*, 2009; NHBC, 2013). Ventilation and the consequences of increased mechanisation of ventilation on building energy are discussed in more depth in Chapter 7.

2.4.6 Alternative design strategies: passive solar design (PSD)

Energy-efficiency design strategies, such as Passivhaus, which use a fabric first approach combined with mechanised ventilation, improve the energy performance of a building without necessitating significant changes to design or construction practices. This provides a 'one-size-fits-all' approach, replicable in any given location, and fits well into modern mass production. However, a building's operational energy performance is also related in part to both the micro climate and the physical aspects of the site that a building occupies and the building's built form (Brown and DeKay, 2001; Nicol and Roaf, 2007). These influence how much energy a building will need, and over what time period, during the course of a year. An alternative approach would, therefore, begin with the unique aspects of a site and design a building tailored to that context.

Passive Solar Design (PSD) principles consider the particular aspects of a site that contribute to a site's micro climate. These include the sun's path over the year, the prevailing wind direction and the surrounding landscape (trees, buildings, slopes, etc.). This context is used to suggest potential design strategies that can best exploit the available solar energy in that environment for providing heating, cooling, daylighting and ventilation, without the need for energy-hungry mechanical systems such as mechanical ventilation (Brown and DeKay, 2001) (Table 2.1).

Using careful planning and design of architectural elements (including glazing, shading, thermal mass and internal room layout) PSD optimises solar gain by converting available sunlight into usable heat or air movement (Crosbie, 1998). Passive solar design has been estimated to reduce the energy required to heat a home by up to 17 per cent (Spanos *et al.*, 2005). However, PSD strategies are not a substitute for insulation and airtightness; an uninsulated leaky building will never be an energy-efficient one. Rather, PSD can only be effective when used in conjunction with these two essential strategies of low-energy building design (Crosbie 1998).

Examples of this approach include architects Robert and Brenda Vales' Autonomous House in Nottingham, constructed in 1993 (Vale and Vale, 2010) and the Oxford Ecohouse constructed in 1994 (Roaf, 2007). The Autonomous House is an off-grid low-carbon home constructed to high thermal standards that uses solar technologies, including solar thermal and PV, to meet energy demands. The Oxford Ecohouse similarly applies PSD strategies using a solar sunspace in conjunction with thermal mass and super-insulation. Interestingly, the home also has an automated system to operate the home to exploit the available solar resource fully and is an example of a hybrid of passive and active systems. The Oxford Ecohouse, which has been subject to many years of performance monitoring, has an energy demand of just 27 kWh/m²·yr without recourse to other mechanical systems (Roaf *et al.*, 2007).

Energy efficiency through good design, the application of high levels of insulation and airtightness can radically reduce the energy demand of buildings. While easier to execute in new buildings, these relatively simple strategies are equally applicable to existing buildings. However, they come at a cost as they require more time to design, more materials of higher quality and greater attention to detail.

2.4.7 Electricity: appliances and lighting

As discussed in Chapter 1, the demand for electricity is increasing in many countries, driven mainly by the increase in appliances, particularly ICT and consumer electronics. The energy efficiency of many appliance types has improved significantly over the past 20 years (see Figure 1.24). For example, the energy-efficiency index of cold appliances improved by 44 per cent over 14 years (1993–2007), and the introduction of low-energy lighting has increased efficiency considerably. However, these improvements have not been sufficient to balance increased numbers of appliances owned and used. In some countries, such as the UK, there is the further driver of increasing numbers of households.

Although energy-efficient appliances are now widely available, many households own old, inefficient equipment. A study of UK residential electricity use found a wide variation in the average age of appliances, ranging from 3.8 years for kettles to 8.4 years for fridge freezers (Palmer and Terry, 2014). Only a very small proportion of homes owned the most efficient available appliances. With the most efficient models using at least 20 per cent less energy than that of the most commonly owned, the study estimated that upgrading appliances to more efficient models across the UK could result in a saving of 14 per cent of household electricity demand. However, there is the additional problem that many appliances, such as fridges and televisions, are getting larger over time, which generally results in an increase in energy demand, size often being more relevant than the energy rating (Palmer and Terry, 2014).

Globally, electric lighting accounts for almost 20 per cent of total global electricity production (IEA, 2014), resulting in 1,900 $mtCO_2$ emissions per year (Waide and Tanishima, 2006). Yet 1.6 billion people live without access to electric lighting. As nations such as China and India continue to develop, demand is predicted to be 80 per cent higher by 2030 (Waide and Tanishima, 2006). However, with the adoption of more efficient lighting technologies, plus improving building design to make better use of natural lighting, the energy demand of lighting could be reduced by 40 per cent by 2050 compared to current levels.

The most energy-efficient lighting choice in buildings is daylighting, which can save up to 70 per cent of lighting energy in a specific day-lit area (Waide and Tanishima, 2006). Not only can daylighting reduce energy demand, but day-lit buildings are also attributed with significant benefits, including occupant satisfaction, increased sense of well-being and health, and higher productivity (Waide and Tanishima, 2006). However, daylighting can be technically challenging to use. Daylight is variable with the weather and seasonally, and it can cause glare and contribute significant heat gains that may be unwanted, particularly in well-insulated buildings.

The second option for efficient lighting is to replace inefficient technologies such as incandescent bulbs, which only convert 5 per cent of the electricity to light, the remainder being converted to heat. Fortunately, driven in Europe by the EU Eco-design Directive, incandescent bulbs have been phased out, leading to a substantial growth in CFLs and LEDs. CFLs use approximately 80 per cent less energy for the same light output as an incandescent bulb, and have a relatively long life (10,000 hours).

LED lighting is developing quickly and may supersede CFLs in the future. An LED fitting has very low power consumption (0.1–15 W) and an extremely long life (10,000–50,000 hours), two to three times longer than CFLs. As a relatively new technology, cost is the main barrier to its widespread adoption, but LEDs are becoming more widely available and costs are reducing. LEDs offer a second lighting energy efficiency revolution, but the gains may be less than previous lighting innovations.

Although incandescent bulbs are no longer manufactured in Europe, some retailers and householders have stockpiled them. In a UK study, only 30 per cent of the lightbulbs in the households studied were low energy (Palmer and Terry, 2014). Also, the energy savings from technological advancements such as LED lighting may be negated by the potential for increased consumption of lighting in new and attractive ways (Tsao *et al.*, 2010; Tsao and Waide, 2010). This effect, termed the rebound effect, is discussed further in Section 4.8.3.

Changes in lighting illustrate the interaction between different systems and highlight the need for thinking about buildings in a holistic way. Replacing inefficient lighting with more efficient technologies results in significant reductions in heat gains and, as a direct consequence, will influence the need for heating, cooling and ventilation. This has particular relevance for non-residential buildings, such as offices and retail, which have high internal heat gain from lighting and other appliances such as refrigeration or computing.

In addition to the savings made by reducing the electricity demand for lighting, a further 30–40 per cent saving may be achieved from the reduction in cooling and ventilation requirements. An often-cited 'rule of thumb' is that 1 kWh of air conditioning energy is saved for every 3 kWh of lighting energy saved, although the net energy savings are dependent on climate, building characteristics, use and the local climatic characteristics (Sezgen and Koomey, 2000).

The Eco-Design of Energy-Related Products Directive (Eco-Design Directive) provides a framework for setting minimum environmental performance and energy-efficiency requirements for energy-related products, with new standards recently adopted for air conditioners and comfort fans, dishwashers, washing machines and lighting. The Directive was extended in 2009 to products that influence energy use, such as windows, although in practice it is mainly used to set energy-efficiency performance criteria. Thus there is a future opportunity to steer product design in a more sustainable direction, with, for example, requirements on repairability or upgrading in order to prevent waste. This Directive is supplemented by the revised Energy Labelling Directive and the EU Eco Label Regulation (EC, 2009). However, resultant energy-efficiency improvements in appliances from these Directives have, to some extent, been offset by increasing ownership and use.

2.5 Energy supply: micro-generation

Although this book primarily focuses on energy demand, reducing the carbon emissions of buildings also depends on decarbonising the energy supply. Exploiting renewable energy resources, and deploying low-carbon technologies, involves decreasing our dependence on fossil fuels, as opposed to reducing our overall need for energy. Reinforcing the energy hierarchy framework, it makes sense to focus on energy supply after tackling energy demand because the demand will, in theory, be smaller and therefore decarbonisation easier to achieve.

We note that energy from low-carbon technologies and from renewable energy sources occurs at different scales. Here, we are concerned with energy supply changes in the context of supplying energy to individual buildings and/or small communities ('micro-generation'). As such, it can be considered to be reducing a building's demand for energy from the carbon-intense electricity and gas grid. Therefore, although energy supply is generally beyond the scope of this book, this short section introduces some of the principal renewable energy and low-carbon technologies currently being deployed at building and community specific scales.

Renewable energy, being derived from renewable resources, has net zero carbon emissions during operation, although energy is required for other parts of their lifecycle (e.g. during the manufacture and disposal of the technologies). Renewable energy can provide heat (e.g. solar thermal, biomass, biofuels) or electricity (e.g. solar photovoltaics, wind turbines, hydroelectricity). In contrast, low-carbon technologies use energy resources derived from conventional fuels, such as nuclear, or use technologies to improve efficiency, such as combined heat and power (CHP) and heat pumps (Table 2.1).

Wind, hydro, solar photovoltaics (PV) and solar thermal have come to the fore as relatively mature technologies offering significant energy and carbon reductions. Solar PV and solar thermal, in particular, have gained prominence within the built environment as the technologies that fit well within current construction practice and regulatory frameworks in countries such as the UK and Germany (Kierstead, 2008).

As many renewable energy resources rely on the energy from the sun, either directly or indirectly, they are dependent upon the availability of this resource; solar energy, being limited seasonally, is discontinuous and highly variable. Furthermore, the deployment of low-carbon technologies is constrained by many factors, including physical (e.g. topography, landscape) and social factors (e.g. planning policy, local acceptance). As a consequence, low-carbon technologies are typically deployed as supplementary systems to a more consistent principal system which will typically be derived from fossil fuels. For example, in a home in the UK a solar thermal system may supplement an electric immersion or gas-fired boiler system. Other low-carbon technologies that can exploit either renewable energy or conventional fossil fuels are those that generate energy (heat and/or electricity) more efficiently and include heat pumps and combined heat and power (CHP).

In the following sections we explore four micro-generation technologies: solar hot water (SHW), PV, heat pumps (HPs) and biomass. Wind power is excluded because, although small-scale roof-mounted turbines are available, they have been shown to be inefficient, generating very small amounts of electricity (Peacock *et al.*, 2008).

2.5.1 Solar hot water

SHW systems convert sunlight into hot water by either directly heating water or, more usually, via a closed heat exchanger. In homes, the use of SHW systems can provide an average 60 per cent of a household's hot water demand (EST, 2011). This can result in significant carbon savings, depending upon the fuels being displaced. For example, if gas is being replaced, an average of 230 $kgCO_2$ per year can be saved, while if an electric immersion heater is replaced, an average of 510 $kgCO_2$ per year can be saved (EST, 2011). SHW systems also require electricity for pumps and controls of between 10 and 180 kWh per year (BRE, 2009; EST, 2011), resulting in carbon emissions of 4.6–88 $kgCO_2$ per year, but this is typically a small fraction of the total savings from SHW systems and is not thought to be a critical issue.

However, the contribution that an SHW system can make to a household's overall hot water heating demand is highly variable, ranging from 9 to 98 per cent (BRE, 2009; EST, 2011). The conditions of use have been shown to be a critical factor in determining the amount of hot water derived from the SHW system (BRE, 2009; EST, 2011). These include the volume of hot water demand (higher hot water demand enables greater solar contribution), the timing of input from subsidiary heating systems ('topping up' at the end of the day rather than beginning) and temperature (higher temperatures require significantly more 'top-up' from subsidiary systems and increased heat loss from the cylinder).

2.5.2 Photovoltaics

PV arrays vary in size and thus generation capacity. PV arrays are modular, comprising of a number of PV modules connected together in series. The power rating of individual modules varies by manufacturer and model. The PV array's size is usually given in kW peak (kWp), which is the number of panels multiplied by their individual module power rating.

Not all the power produced by a PV array is available for use, as a result of efficiency losses, including those associated with the conversion from direct current (DC) to alternating current (AC). An early study of 170 1–5 kWp grid-connected PV systems in Germany found that system losses ranged from 10 to 16 per cent (Decker and Jahn, 1997), although more recently technical improvements have decreased the losses to be closer to 10 per cent (Ayompe *et al.*, 2011). The climatic conditions (e.g. cloudiness, temperature) also influence the conversion efficiency (i.e. the amount of available sunlight converted to DC current) of the modules (Ayompe *et al.*, 2011; So *et al.*, 2007). The conversion efficiency also declines over time due to photon degradation, severe discoloration, delamination, cracking of cover glass, splitting of backsheets, wiring degradation and junction box failure (Dunlop and Halton, 2006). However, the extent of these issues is not likely to become known until these systems have been in place for a number of years.

The annual electricity output from a PV array fluctuates from year to year, depending upon climatic factors. Studies on PV performance have found yields to be widely variable. For example, studies on PV systems in Germany have found a range in system yields of 400–1,030 kWh per year, averaging 885 kWh per year (Decker and Jahn, 1997; Jahn and Nasse, 2004).

Although PV output can be accurately estimated (Bahaj and James, 2007), there are few studies of grid-connected PV systems that measure, or estimate, the proportion of PV-generated electricity that is utilised directly within the building, compared with that exported to the grid. The studies that are available indicate a very wide range, between 20 and 73 per cent used on-site (Erge *et al.*, 2001; Bahaj and James, 2007). The proportion used in homes is determined by occupation patterns, with households that change the timing of activities such as washing and cooking (termed load shifting) to exploit the PV-generated electricity using the greatest proportion of PV electricity (Bahaj and James, 2007). This study of nine UK domestic PV systems concluded that PV can provide a contribution towards the annual electricity demand of a household, but this will be limited if there is not a concomitant load shifting of the demand.

2.5.3 Heat pumps

Heat pumps, using the same process as found in a fridge or air conditioning unit, move low-grade heat from a source (e.g. air, ground or water) to a heat sink (i.e. heating and hot water). Air source heat pumps (ASHPs) and ground source heat pumps (GSHPs) are currently the most widely installed in the UK. The process requires electricity for fans, pumps and controls, but for every unit of electricity required, approximately 2–4 units of useful heat are produced. This ratio is called the coefficient of performance (CoP) and is measured as the heating output (in kW) divided by the total electricity consumed by the heat pump system (also in kW). However, the CoP is dynamic, varying with the temperature difference between input and output. The heat source will vary over the heating season as it is depleted, and the demands placed upon it in terms of output will also vary. CoP is, therefore, both a measure of the effectiveness of the heat pump system itself and a measure of the conditions in which it is used.

In a recent study of 83 installed ASHP and GSHP systems, the Energy Saving Trust (EST, 2010) found 87 per cent of the monitored systems underperformed, with GSHPs having CoPs that ranged between 1.3 and 3.6. The report cited a number of contributory factors relating to design (including the sizing of pumps, ground loops, hot water cylinders and heat emitter area), system installation (including poor insulation, commissioning and incorrect temperature set up) and occupant behaviour. Despite the poor performance results, the report concluded that heat pumps reduce CO_2 emissions when compared with conventional gas heating technologies, but are most effective when displacing conventional electric or oil-fired heating systems.

2.5.4 Biomass

Log-burning stoves are fairly common, often performing as a secondary heat source and providing a focal point in a living room. More recently there has been an introduction in the UK of automated biomass boilers that combust wood chips and pellets. Although these are relatively uncommon, they are increasing in the UK due to the introduction of the Renewable Heat Incentive that subsidises biomass when it is used as the primary heat source (DECC, 2011). Although biomass stoves and boilers save fuel costs, they have a high capital cost, with automated biomass boilers being particularly expensive (costing £9,000–£21,000 to purchase and install) and requiring a dry storage area for the biomass pellets or chips (EST, 2014). Furthermore, there is some concern about the health impacts of wood combustion (Naeher *et al*, 2007), particularly respiratory illnesses from particulate emissions, so they are often considered more suitable in rural rather than urban areas.

2.6 Reflecting upon the energy hierarchy's technological focus

The underlying approach intrinsic to the energy hierarchy relies upon a technological solution to deliver energy demand reductions and lower carbon emissions. Indeed, the challenge of addressing energy conservation is increasingly being regarded as a technological one, and is the only stage of the hierarchy that considers how technologies are used (rather than just trying to target uptake and installation of technologies). From this perspective, technological systems, and the interactions of users with these systems, appear to be straightforward and linear. In other words, a physiologically determined need is met by the provision of a technology and it is the technology that determines the amount of an energy service required to meet that need. For example, the need to feel warm is met by a heating system and the energy use of that system can be 'optimised' with the 'right' control systems and 'right' user behaviour, delivering the 'right' amount of heat at the 'right' time.

However, in reality, the processes that underpin energy demand are anything but linear. Energy demand is messy and complicated. It is the expression of many interdependent actors ranging from local and national governments, utility companies, other built environment professionals, to the building occupants themselves (Chappells and Shove, 2003). Therefore, using the previous heating example, it is assumed that space heating's carbon emissions are a function of the technical characteristics of a building (e.g. type, levels of thermal insulation, air permeability), the heating technology employed (e.g. efficiency of combustion, levels of control) and the fuels used (e.g. gas, oil, wood). However, by acknowledging that there are broader influences in play, one can start to appreciate that technologies may be appropriated and used in different, and perhaps even unexpected, ways. Consequently, the energy hierarchy's technological focus could be sending a (inappropriate) message that all that is required is to pick the 'right' type of technology, which could 'guarantee' energy demand reduction, irrespective of how those technologies may actually be used in practice. This sort of broader perspective is a position that we regularly return to throughout this book.

2.7 Case studies

2.7.1 Case study 1: Student Switch-Off campaign

Rationale

The Student Switch-Off campaign, first piloted at the University of East Anglia in 2007, is a not-for-profit international campaign encouraging action on climate change in which students are encouraged to take simple steps to save energy and so help participating universities to reduce their energy demand. We include this as a case study of an energy conservation initiative that focuses exclusively on behaviour change.

Project background

- Location: international campaign with 54 participating universities in the UK, Sweden, Cyprus, Greece, Lithuania and Canada.
- Building type: university student halls of residence.
- Company: UK-based not-for-profit company run by the UK's National Union of Students with sponsorship from Ben and Jerry's and co-funded by the Intelligent Energy Europe Programme of the European Union.
- Approach: energy demand reduction through behaviour change by incentivising students to reduce their energy demand by offering financial and other rewards in an annual competition.

Approach

There is generally no incentive for students in university residences to conserve energy. Typically, energy costs are included in fixed accommodation costs. Consequently, there is a substantial opportunity to reduce energy demand in university residences. Furthermore, it is also assumed that instituting energy conservation behaviour changes at a major milestone in a young person's life may set up energy conservation norms that will be embedded for life.

The Student Switch-Off campaign encourages students to take simple behaviour changes to reduce energy demand, including wrapping up warm, switching off lights and using lids on saucepans. At participating universities students compete to see which hall of residence can achieve the greatest reductions in electricity demand over the academic year. Individual student 'EcoPower Rangers' are recruited and trained, learning how to motivate their fellow students to join in with the energy conservation efforts. At the end of the year, rewards are awarded to the winning halls. Innovative methods, including social media

(Facebook, YouTube, Twitter, etc.), peer-to-peer engagement and student-focused incentives and activities such as photo competitions and quizzes are used to engage students. Incentives, which include luxury foodstuffs and tickets to events, are awarded for energy-efficient behaviour at individual and communal level. The methods employed build on existing social relationships, rivalries and communities. The results are measured via electricity meters and feedback is provided to each hall of residence throughout the year.

Findings

The Student Switch-Off campaign resulted in an overall 6 per cent reduction in electricity demand across all participating halls of residence in the 2013–2014 academic year. This was estimated to reduce energy costs by £261,000 and carbon emission by 1,606 tonnes of CO_2. The campaign monitors the results over time as the campaign has spread to more universities (Table 2.6). The results show a reduction in electricity demand of 7 per cent between 2007 and 2014.

As discussed in Section 2.3.1, initiatives that rely on awareness raising to institute behaviour change have been criticised for lack of success. In 2011 the Student Switch-Off campaign undertook a survey of participating universities to investigate the wider impacts of participating in the campaign. The survey headline results ($n = 636$ respondents) noted that:

- 88 per cent had taken action directly as a result of the Student Switch-Off campaign;
- 84 per cent had increased personal awareness of energy conservation;
- 93 per cent stated they were likely or very likely to continue the energy-saving behaviour;
- 74 per cent had encouraged their friends to save energy.

The results suggest that the underlying aim of the campaign, to institute an awareness of energy demand and lifetime behaviour changes in participants at a period of change in

Table 2.6 Mean annual results of Student Switch-Off campaign (2006–2014)

Year	No. of universities	No. of students signed up	Electricity reduction (%)	CO_2 saved (tonnes)	Money saved (£)
2006–2007	1	130	10	90	19,000
2007–2008	7	2,800	8.9	550	100,000+
2008–2009	11	4,980	9.3	1,295	218,000
2009–2010	33	12,052	6.9	2,100	337,000
2010–2011	37	15,351	6.9	1,522	232,000
2011–2012	43	19,430	5.7	1,405	220,000
2012–2013	54	22,715	5.3	1,478	219,000
2013–2014	54	26,812	5.9	1,606	261,000

Source: Student Switch-Off (2014).

their lives, may be achieved. However, the sample of respondents is small and self-selecting; consequently the results should be viewed with caution. Furthermore, it is not known whether participants continue this behaviour once they leave halls and move into homes of their own, or if energy conservation behaviours, instituted by the campaign within the halls of residence, would become normalised so would continue without the continuing intervention of the campaign.

Key messages

- Using a targeted approach tailored to a specific audience with appropriate incentives resulted in energy conservation-related behaviour change (an average 7 per cent energy saving per year).
- The results suggest that the underlying aim of the campaign, to institute an awareness of energy demand and lifetime behaviour changes in participants at a period of change in their lives, may be achieved. However, energy conservation behaviour change may only occur for as long as the campaign is visible rather than being institutionalised behaviour.

2.7.2 Case study 2: Thomas Paine Study Centre, University of East Anglia, Norwich, UK

Rationale

This case study demonstrates the application of the fabric first approach to constructing an energy-efficient building that includes substantial thermal mass and high levels of airtightness and thermal insulation.

Project background

- Location: University of East Anglia, Norwich, UK
- Building type: education
- Tenure: owner-occupier
- Year of construction: 2010
- Size: 4,301 m^2
- Construction approach: fabric first using Swedish TermoDeck system, which is a hollowcore reinforced concrete floor slab that provides high thermal mass and ventilation in combination with high-efficiency heat-exchange units and a sophisticated energy management system.

Approach

The Thomas Paine Study Centre is one of the University of East Anglia's most recent in a series of evolving low-energy buildings using the Swedish Termodeck system that began in 1995 with the Elizabeth Fry Building (see Section 4.10.1 for details of the Elizabeth Fry Building).

The building has super-insulated walls and floor, with triple-glazed windows and very high airtightness of 4.81–3.94 m^3/h/m^2 at 50 Pa (1.7–1.9 ach) (Table 2.7).

Table 2.7 TPSC, UEA building fabric performance

Building element	U-value (W/m²·K)
Floors	1.3
Walls	0.2
Roof	0.13
Glazing	0.13
Designed air permeability	<5 m³/h/m²@ 50Pa

TermoDeck is a structural hollowcore floor slab through which warmed or cooled fresh air is distributed through the building by the provided ventilation. Air is driven by fans through the hollowcore floor slabs at very low speeds, allowing the high-mass slabs to behave as passive heat-exchange elements, thus ensuring stable temperatures with low energy use. Slab temperatures are monitored and controlled to remain within pre-defined set points. If slab temperatures stray from these set points then warmer or cooler air is provided. In most cases this is through taking advantage of external air temperatures and passing air over high-efficiency heat exchanges. Alternatively, heating or cooling of the incoming air is provided by the district heating and cooling mains in the university. For example, in summer the slabs absorb the heat during the day and the building is cooled by night time venting, whereby cooler external air is passed through the slabs to cool them. Conversely, during winter fresh air is heated via the heat exchanger (with supplementary heating from the mains if required) before the building is occupied. The heating is switched off when the slabs are within set parameters. The slabs then release the stored heat, warming incoming air and radiating heat into the occupied spaces.

The building also has a complex energy management system. It is subdivided into zones, each separately controlled for fresh air supply and temperature. Zones are arranged according to different functions (lecture theatre, seminar rooms, offices, WCs and service areas). Each zone has different demands, and zoning permits increased controllability and greater energy savings to be achieved. The building is managed by approximately 800 sensors and controls, which feed data into building energy management system (BEMS) control software, which adjusts fan speeds, dampers, heating and cooling input as required.

Findings

This fabric first approach produced a remarkably low-energy building that far exceeds the UK's current building regulations. Heat energy consumption was 34.8 kWh/m²·yr, cooling 9.0 kWh/m²·yr and electricity 74.9 kWh/m²·yr.

Key to this was a building performance evaluation process during which monitoring was used to refine the building's energy consumption. It was noted with the Elizabeth Fry Building that the built performance, although good, was not as good as other buildings constructed in the same way (EBPP, 1998). A programme of monitoring enabled the tweaking of control systems, which reduced energy demand and improved performance. Subsequently, all of the UEA's low-energy buildings, of which the TPSC is no exception, have a control strategy for building management that was developed over a period of settling-in. This enables monitoring, management and refinement of the automated controls, which in turn results in the performance of TPSC being better than its predecessors (Ingham, 2012).

Key messages

- An innovative yet simple energy-efficient ventilation system, in which the building's hollowcore structure provides both ventilation and thermal mass, used in combination with very low airtightness, produced a building with low energy demand.
- A sophisticated energy management system and automation was integral to the design and operation of the services, including ventilation, heat recovery, lighting, etc.
- A lengthy period of settling-in, with extensive monitoring, management and refinement of systems was essential in reducing the building's initial energy consumption to close to its designed performance.

2.8 Summary

There is a significant potential for reducing the energy demand from buildings using existing and mature technologies together with design approaches. The energy hierarchy's sequential priorities, to reduce energy demand before addressing energy supply, provide a clear logical framework for reducing and decarbonising energy demand from the built environment, which is relevant to both new and existing buildings. Energy conservation, using less energy or going without a service to reduce demand, predominantly requires a change in behaviour, such as switching off lights and turning off equipment and appliances when not in use. It also includes aids to reducing demand such as the provision of information on energy use via a visual display meter that aims to make energy more visible to the user. In comparison, energy efficiency, using less energy without reducing the level of service, has a technical approach, such as improving levels of insulation or installing an efficient boiler. Many seemingly technical approaches, however, also have social aspects, such as the need for high-quality workmanship and supervision in order to achieve airtight buildings.

Within the hierarchy framework there are numerous technical and behavioural strategies that can be employed, ranging from design, detailing, thermally efficient materials and smart technology to behaviour change and controls. However, the use and effectiveness of these strategies is constrained by how we design, construct and manage our buildings. The difficulty of achieving large-scale behaviour change may lead to increased automation of building controls and the establishment of smart homes, which may serve the additional purpose of supporting our ageing population as well as reducing energy demand. However, the intervention of a smart home approach is likely to be as much for the use and expediency of the energy supplier as for the occupant.

The construction of well-insulated, thermally 'tight' buildings can paradoxically lead to high humidity and poor air quality, causing damp, mould and health problems. Technical solutions can include passive ventilation, but are more likely to be mechanical ventilation systems, usually incorporating heat recovery.

Significant progress has been made in the energy efficiency of appliances and lighting, but in developed countries much of that achievement has been negated by a substantial increase in the number of appliances per household, and the number of households. In developing countries the increase in electrification, lighting and use of appliances will have substantial implications for energy demand and carbon emissions in the future.

References

Abrahamse, W., Steg, L., Vlek, C. and Rothengatter, T. (2005) A review of intervention studies aimed at household energy conservation. *Journal of Environmental Psychology*, 25(3), 273–291.

Ayompe, L.M., Duffy, A., McCormack, S.J. and Conlon, M. (2011) Measured performance of a 1.72 kW rooftop grid connected photovoltaic system in Ireland. *Energy Conversion and Management*, 52, 816–825.

Bahaj, A.S. and James, P.A.B. (2007) Urban energy generation: The added value of photovoltaics in social housing. *Renewable and Sustainable Energy Reviews*, 11, 2121–2136.

Beggs, C. (2010) *Energy: Management, Supply and Conservation*, Routledge, London.

Boait, P.J. and Rylatt, R.M. (2010) A method for fully automatic operation of domestic heating. *Energy and Buildings*, 42, 11–16.

Boyle, G. (2012) *Renewable Energy: Power for a Sustainable Future*, Open University Press, Milton Keynes.

BRE (2009) Clearline solar thermal test report: average household simulation. Viridian solar client report number 251175, Building Research Establishment.

Brown, G. and Dekay, M. (2001) *Sun, Wind and Light: Architectural Design Strategies*, John Wiley and Sons Ltd, Chichester.

Chappells, H. and Shove, E. (2003) The environment and the home. Draft paper for Environment and Human Behaviour Seminar, Policy Studies Institute, London, 23 June.

CIBSE (2006) *CIBSE Guide A: Environmental Design*, CIBSE, London.

Combe, N., Harrison, D., Dong, H., Craig, S. and Gill, Z. (2010) Assessing the number of users who are excluded by domestic heating controls. *International Journal of Sustainable Engineering*, 4, 84–92.

CPA (2014) *Update on Preparation for Smart Metering. Twelfth Report of Session 2014–15. House of Commons Committee of Public Accounts*, HMSO, London.

Crosbie, M. (1998) *The Passive Solar Design and Construction Handbook*, John Wiley and Sons, New York.

Crump, D., Dengel, A. and Swainson, M. (2009) *Indoor Air Quality in Highly Energy Efficient Homes: A Review*, National House Building Council, Milton Keynes.

Darby, S. (2006) The effectiveness of feedback on energy consumption, a review for DEFRA of the literature on metering, billing and direct displays. Environmental Change Institute, University of Oxford. www.eci.ox.ac.uk/research/energy/downloads/smart-metering-report.pdf, accessed 26 May 2015.

Darby, S. (2010) Smart metering: What potential for householder engagement? *Building Research Information*, 38(5), 442–457.

DCLG (2013) *Approved Document L: Conservation of Fuel and Power*, DCLG, London.

DECC (2012) *Energy Efficiency Strategy: The Energy Efficiency Opportunity in the UK*, HMSO, London.

DECC (2011) Renewable heat incentive. Department of Energy and Climate Change.

Decker, B. and Jahn, U. (1997) Performance of 170 grid connected PV plants in Northern Germany: Analysis of yields and optimization potentials. *Solar Energy*, 59, 127–133.

DEFRA (2011) *Guidance on Applying the Waste Hierarchy*, HSMO, London.

Dunlop, E.D. and Halton, D. (2006) The performance of crystalline silicon photovoltaic solar modules after 22 years of continuous outdoor exposure. *Progress in Photovoltaics: Research and Applications*, 14(1), 53–64.

EBPP (1998) The Elizabeth Fry Building, University of East Anglia, new practice final report 106. www.termodeck.com/Filer/pdf/BRE_EFRY_REPORT.pdf, accessed 9 June 2015.

EC (2009) Regulation (EC) No 66/2010 of the European Parliament and of the Council of 25 November 2009 on the EU Ecolabel. http://eur-lex.europa.eu/legal-content/EN/TXT/?uri=CELEX:32010R0066, accessed 9 June 2015.

Elliott, D. (2013) *Renewables: A Review of Sustainable Energy Supply Options*, IOP Publishing Ltd, Bristol.

Erge, T., Hoffmann, V.U. and Kiefer, K. (2001) The German experience with grid-connected PV-systems. *Solar Energy*, 70, 479–487.

EST (2015) Domestic home insulation. www.energysavingtrust.org.uk/domestic/content/home-insulation, accessed 9 June 2015.

EST (2014) Biomass. www.energysavingtrust.org.uk/domestic/content/biomass, accessed 19 August 2015.

EST (2012) Powering the nation: household electricity-using habits revealed. www.energysavingtrust.org.uk/sites/default/files/reports/PoweringthenationreportCO332.pdf, accessed 26 May 2015.

EST (2011) Here comes the sun: a field trial of solar water heating systems. http://tools.energysavingtrust.org.uk/Publications2/Generating-energy/Field-trial-reports/Here-comes-the-sun-a-field-trial-of-solar-water-heating-systems, accessed 26 May 2015.

EST (2010) Getting warmer: a field trial of heat pumps. www.heatpumps.org.uk/PdfFiles/TheEnergySavingTrust-GettingWarmerAFieldTrialOfHeatPumps.pdf, accessed 26 May 2015.

GLA (2014) Energy Planning, Greater London Authority guidance on preparing energy assessments. www.london.gov.uk/priorities/planning/strategic-planning-applications/preplanning-application-meeting-service/energy-planning-gla-guidance-on-preparing-energy-assessments, accessed 26 May 2015.

Hargreaves, T., Nye, M. and Burgess, J. (2013) Keeping energy visible? Exploring how householders interact with feedback from smart energy monitors in the longer term. *Energy Policy*, 52, 126–134.

Huovila, P. (2007) *Buildings and Climate Change: Status, Challenges, and Opportunities*, UNEP/Earthprint, Paris.

IEA (2014) About lighting. www.iea.org/topics/energyefficiency/subtopics/lighting, accessed 9 June 2015.

Ingham, M. (2012) Thomas Paine Study Centre: How it works. University of East Anglia, Low Carbon Innovation Centre and Build with CaRe.

Jahn, U. and Nasse, W. (2004) Operational performance of grid-connected PV systems on buildings in Germany. *Progress in Photovoltaics: Research and Applications*, 12, 441–448.

Johnston, D., Miles Shenton, D., Wingfield, J. and Bell, M. (2004) Airtightness of UK dwellings: Some recent measurements. *Proceedings of the Construction and Building Research Conference, London*. Edited by Ellis, R. and Bell, M.

Keirstead, J. (2008) What changes, if any, would increased levels of low-carbon decentralised energy have on the built environment? *Energy Policy*, 36, 4518–4521.

Keirstead, J. (2007) Behavioural responses to photovoltaic systems in the UK domestic sector. *Energy Policy*, 35, 4128–4141.

Liao, Z., Swainson, M. and Dexter, A.L. (2005) On the control of heating systems in the UK. *Building and Environment*, 40, 343–351.

Lowe, R., Johnston, D. and Bell, M. (1997) Airtightness in UK dwellings: a review of some recent measurements. In *Proceedings of the 2nd International Conference on Buildings and the Environment*, 9–12 June, Paris.

Metz, B., Davidson, O.R., Bosch, P.R., Dave, R. and Meyer, L.A. (eds) (2007) *Climate Change 2007: Mitigation of Climate Change: Contribution of Working Group III to the Fourth Assessment Report of the Intergovernmental Panel on Climate Change*, Cambridge University Press, Cambridge.

Monahan, J. (2013) Housing and carbon reduction: can mainstream 'eco-housing' deliver on its low carbon promises? PhD thesis, University of East Anglia, Norwich, UK.

Naeher, L.P., Brauer, M., Lipsett, M., Zelikoff, J.T., Simpson, C.D., Kownig, J.Q. and Smith, K.R. (2007) Woodsmoke health effects: A review. *Inhalation Toxicology*, 19(1), 67–106.

NHBC (2013) Mechanical ventilation with heat recovery in new homes final report. The Ventilation and Indoor Air Quality Task Group NHBC Foundation. www.nhbcfoundation.org/Portals/0/Viaq-Final-Report-July2013.pdf, accessed 26 May 2015.

Nicol, F. and Roaf, S. (2007) Progress on passive cooling: Adaptive thermal comfort and passive architecture. In: Santamouris, M. (ed.), *Advances in Passive Cooling*, Earthscan, London.

Palansky, P. and Kupzog, F. (2013) Smart grids. *Annual Review of Environment and Resources*, 38(1), 201–226.

Palmer, P. and Terry, N. (2014) How much energy could be saved by making small changes to everyday household behaviours? Annex Spreadsheet. www.gov.uk/government/publications/how-much-energy-could-be-saved-by-making-small-changes-to-everyday-household-behaviours, accessed 9 June 2015.

Palmer, P., Terry, N. and Pope, P. (2012) How much energy can be saved by making small changes to everyday household behaviours? Department of Energy and Climate Change. www.gov.uk/government/uploads/system/uploads/attachment_data/file/128720/6923-how-much-energy-could-be-saved-by-making-small-cha.pdf, accessed 26 May 2015.

Peacock, A.D., Jenkins, D., Ahadzi, M., Berry, A. and Turan, S. (2008) Micro wind turbines in the UK domestic sector. *Energy and Buildings*, 40(7), 1324–1333.

Phillips, T., Rogers, P. and Smith, N. (2011) Ageing and airtightness: How dwelling air permeability changes over time. NHBC Foundation. http://products.ihs.com/cis/Doc.aspx?AuthCode=&DocNum=296303, accessed 26 May 2015.

Rashidi, P. and Cook, D.J. (2009) Keeping the resident in the loop: Adapting the smart home to the user. *IEEE Transactions on Systems, Man, and Cybernetics*, 39(5), 949–959.

Roaf, S. (2007) *Ecohouse*, Routledge, London.

Roaf, S., Thomas, S. and Fuentes, M.T.S. (2007) *Ecohouse*. 3rd edition. Routledge, London.

Sezgen, O. and Koomey, J.G. (2000) Interactions between lighting and space conditioning energy use in US commercial buildings, *Energy*, 25(8), 793–805.

So, J.H., Jung, Y.S., Yu, G.J., Choi, J.Y., and Choi, J.H. (2007) Performance results and analysis of 3 kW grid-connected PV systems. *Renewable Energy*, 32, 1858–1872.

Spanos, I., Simons, M. and Holmes, K. (2005) Cost savings by application of passive solar heating. *Structural Survey*, 23, 111–130.

Stephen, R. (2000) *Airtightness in UK Dwellings*, BRE, Watford.

Stevenson, F., Carmona-Andreu, I. and Hancock, M. (2013) The usability of control interfaces in low-carbon housing. *Architectural Science Review*, 56, 70–82.

Student Switch-Off (2014) Student Switch-off campaign. www.studentswitchoff.org, accessed 9 June 2015.

Tsao, J.Y. and Waide, P. (2010) The world's appetite for light: Empirical data and trends spanning three centuries and six continents, *LEUKOS*, 6(4), 259–281.

Tsao, J.Y., Saunders, H.D., Creighton, J.R., Coltrin, M.E. and Simmons, J.A. (2010) Solid-state lighting: An energy-economics perspective. *Journal of Physics D: Applied Physics*, 43(35), doi:10.1088/0022-3727/43/35/354001.

UNEP (2007) *Buildings and Climate Change: Status, Challenges and Opportunities*, United Nations Environment Programme, Sustainable Buildings and Construction Initiative, Paris.

Vale, B. and Vale, R. (2010) Domestic energy use, lifestyles and POE: Past lessons for current problems. *Building Research and Information*, 38, 578–588.

van Dam, S.S., Bakker, C.A. and Van Hal, J.D.M. (2010) Home energy monitors: Impact over the medium-term. *Building Research and Information*, 38, 458–469.

Waide, P. and Tanishima, S. (2006). *Light's Labour's Lost: Policies for Energy-efficient Lighting*, OECD Publishing, Paris.

Ward, T.I. (2006) *Assessing the Effects of Thermal Bridging at Junctions and Around Openings*, BRE, Watford.

Williams, J. and Tipper, H.A. (2013) *Low Carbon Behaviour Change: The £300 Million Opportunity*, The Carbon Trust, London.

Wilson, C., Hargreaves, T. and Hauxwell-Baldwin, R. (2015) Smart homes and their users: A systematic analysis and key challenges. *Personal and Ubiquitous Computing*, 19(2), 463–476.

3 Lifecycle energy of buildings

3.1 Introduction

When we consider energy in the built environment, the focus is almost exclusively on the energy associated with the operational phase of a building's lifecycle. Reducing operational energy is, and will continue to be, an important goal and a significant focus for policy. A substantial amount of energy, however, is also used in the construction, refurbishment and eventual demolition (or deconstruction) of a building. The extraction, processing, manufacture, transportation, use and eventual disposal of a material, component or product used in the construction of a building requires energy and produces many environmental impacts, including carbon emissions. These impacts, which are called the hidden (or embodied) environmental burdens, are not insignificant, but are often ignored. For example, the UK Department for Business Innovation and Skills (BIS, 2010) estimated the UK construction sector to be responsible for 16 per cent of the UK's total carbon emissions. This was related to materials, products, transport and the construction process. Significantly, the report found the construction industry has the ability to influence (directly and indirectly) nearly 300 $mtCO_2$ and clearly represents an untapped opportunity for achieving the UK's climate change targets.

Operational energy is currently by far the largest proportion of lifetime energy for conventional buildings (Sartori and Hestnes, 2007), typically accounting for 80–90 per cent, with the embodied energy accounting for 10–20 per cent of total lifetime energy (Ramesh *et al.*, 2010). Another way of expressing embodied energy is to calculate how many years of operational energy the embodied energy is equal to. For example, embodied energy has been found to be equivalent to 7.8 years of operational energy for offices and 7.5 years for homes (Yung *et al.*, 2013). However, in low-energy buildings the relative importance of embodied energy and operational energy changes (Huberman and Pearlmutter, 2008; Winther and Hestnes, 1999). When comparing a low-energy home with a conventional one, Thormark (2002) found the proportion of embodied energy to be equivalent to 37.5 years of operational energy, much greater than a conventional home.

In low-energy buildings, the proportion of embodied energy may increase to 60 per cent or more of the whole lifecycle energy (Blengini and Di Carlo, 2010). This difference between conventional and low-energy buildings can be attributed to a substitution effect (or trade-off) that occurs because additional energy is often required for the manufacture of additional materials, such as greater amounts of insulation, used to achieve increased fabric performance (Winther and Hestnes, 1999; Thormark, 2002). This is not to suggest that embodied energy and operational energy are competitors, but that the desired outcome of optimising operational energy demand can come at a cost to the energy embodied in the construction.

Optimising fabric energy performance may not necessarily result in significantly higher embodied energy. For instance, the Passive House Institute (PHI) found the increased embodied energy needed to meet the improved fabric energy performance required to achieve Passivhaus standards is repaid in a very short time, potentially less than a year (Feist, 1997; Mossman *et al.*, 2005). Further, for conventional construction, increased embodied energy can be mitigated to a certain extent. A recent report by the National House Builders Federation suggests that embodied impacts required to meet higher fabric standards can be reduced by 5–10 per cent without compromising performance (Zabalza-Bribián *et al.*, 2011).

However, there may be a stage in operational energy reduction beyond which it becomes counterproductive. The nearer buildings become to being very low energy or even energy self-sufficient, the greater the need for additional technologies (e.g. mechanical ventilation, solar photovoltaics or solar thermal). These additional technologies will have significant additional embodied energy, as well as having a lower life expectancy compared with thermal insulation. Consequently, the inclusion of such technologies has been shown to increase the environmental impacts of buildings significantly (Winther and Hestnes, 1999; Sartori and Hestnes, 2007). Such an outcome is counterproductive to actions to reduce the energy and environmental impacts of buildings.

Embodied energy is not generally considered when a building is designed, specified and constructed. Yet quantifying the embodied energy of a building enables the identification of 'hotspots', which suggest opportunities for reducing embodied energy and the management of environmental impacts of a building. Hotspots refer to areas for which improved material choice and/or manufacturing practices can reduce energy consumption and remove unnecessary harmful environmental impacts.

In summary, the relative importance of embodied energy increases as buildings become more energy efficient. By focusing on reducing operational energy, only part of the whole picture is considered. Not only does neglecting to account for embodied energy represent a lost opportunity for reducing the environmental impacts of the built environment, it may also have the unintended consequence of increasing the use of energy-hungry and carbon-intensive products, such as plastics in high-performance insulation products. The resulting allocation of resources, when considered over the whole lifetime of a building, may not actually achieve overarching energy and environmental goals.

In this chapter we introduce the principal concepts involved in understanding embodied energy. In particular, we focus on the methodological approaches used to quantify the embodied energy associated with a building over its whole lifetime. We provide suggestions for changes that can be made to reduce embodied energy so as to lower the lifecycle energy and the carbon of buildings. The implications of using 'natural' construction materials are also explored.

3.2 The lifecycle of buildings

Energy is used and emissions occur at all points in the lifecycle stages of a building. The lifecycle of a building is typically viewed linearly, beginning with the extraction of raw materials and ending with demolition and waste management at end-of-life (Figure 3.1). This includes the exploration, finding and extraction of raw materials (e.g. forestry, mining of ores). These are then manufactured into materials, products and components that are transported from the place of manufacture to the construction site. During construction, energy is used in various activities and processes required in constructing the building (e.g. on-site energy used for powering tools, machinery and heavy plant).

Figure 3.1 The lifecycle of a building from cradle to grave.

During the operational life of a building, in addition to the heat and power requirements, there will be recurring embodied energy used in maintaining or replacing items or components at intervals throughout the building's operational life. For example, replacing carpeting every 4–10 years, repainting woodwork and walls every five years or replacing windows every 20 years. Some buildings will also undergo significant retrofitting (see Chapter 5). Finally, at some point a building will reach the end of its useful life and there will be energy demand and carbon emissions associated with the activities of demolishing and disposing of the building and its various materials, products and components. This disposal can include reuse, recycling as a raw material for another manufacturing process, combustion and energy recovery, and/or final disposal to landfill.

Adding together the embodied energy at each stage of a building's lifetime and the operational energy can provide a thorough account of the energy and carbon impacts of that building over its whole lifetime. There have been numerous studies that follow this approach. Meta-analyses of studies of lifecycle energy analyses of buildings have found that the majority of embodied energy occurs prior to construction, during the production of materials and products (61–67 per cent) (Yung *et al.*, 2013). Maintenance was also significant, accounting for 11–21 per cent of lifetime embodied energy, whereas construction and demolition were found to be 7–11 per cent and 11 per cent, respectively. Transportation is typically low, accounting for 1–3 per cent of a building's total embodied energy.

The distinctive stages of a building's lifecycle lends itself to a modular way of thinking about the embodied energy and carbon of a building over its lifetime. This modular approach has been used to define a number of different boundaries that can be applied to the system when measuring and accounting for the energy and carbon embodied in buildings. Typically, these system boundaries are:

- Cradle-to-gate: encompasses raw material extraction (cradle) and product manufacture, up to the point where the material or product reaches the factory gate of the final processing operation. Thus, it excludes transportation to the retailer or purchaser.

- Cradle-to-site: cradle-to-gate plus the transportation of materials and products to the site.
- Cradle-to-construction: from cradle to the incorporation of products into a building on-site; this thereby encompasses all construction activities.
- Cradle-to-grave: cradle to the end of the building's lifetime or study period. Includes the whole lifecycle up to and including end-of-life deconstruction and end-of-life waste management.
- Cradle-to-cradle: cradle-to-grave and beyond, where products are designed so that at end-of-life they can be easily reused or recycled into other products, thus becoming a circular lifecycle.

The end-of-life stage is perhaps the most problematic, particularly as it will occur many tens of years after construction. For example, how will the materials be managed at a given point in the future? At the end-of-life of a building, not all materials will be sent to landfill or combusted to produce energy. In many countries, landfill is becoming less desirable as a method of waste management. For example, the European Waste Directive (EU, 2008) aims to reduce the amount of material sent to landfill and encourage the recycling and reuse of materials. An increasing proportion of materials are separated for recycling, reuse or as raw material in the production of new products. In the future, materials may be regarded less as waste but more as re-usable components, raw materials or feedstock for other processes (known as secondary materials). In such cases, the linear lifecycle shown in Figure 3.1 becomes circular (Figure 3.2). This approach, defined as cradle-to-cradle (BIS, 2010), is not a commonly used system boundary due to the many complications and uncertainties that arise. For example, where would the system boundary be drawn between materials production, the building lifecycle and the production of a new product? And how would the energy and emissions associated with the use of secondary materials be allocated between the original material and the processes used to transform it into a new product?

Because of the problems associated with predicting how buildings will be managed in the future, most embodied energy studies apply a cradle-to-construction boundary. The impacts of maintenance, refurbishment and, in particular, end-of-life management are thus intentionally excluded or ignored. Yet, lifecycle impacts are interdependent, with each stage influencing and being influenced by the other stages. Expanding the boundary to include the whole lifecycle provides not just a complete account, but also a sense of this 'knock on' effect, which is particularly pertinent for early design and specification decisions in both new buildings and retrofitting existing buildings. For example, the selection of a specific building material to reduce heating demand can have significant influence on resources and emissions affecting (1) embodied energy, (2) transport-related impacts, (3) maintenance requirements, (4) total service life of the building and (5) the generation of either demolition waste or useful reusable materials at end-of-life. Furthermore, decisions made with a partial view of the whole lifecycle may lead to erroneous or counterproductive outcomes, resulting in a misallocation of resources. For example, the use of renewable technologies to achieve carbon reductions (compared with fabric interventions) may, in a lifetime context, have greater embodied energy than additional insulation, but have the same impact on reducing primary energy demand.

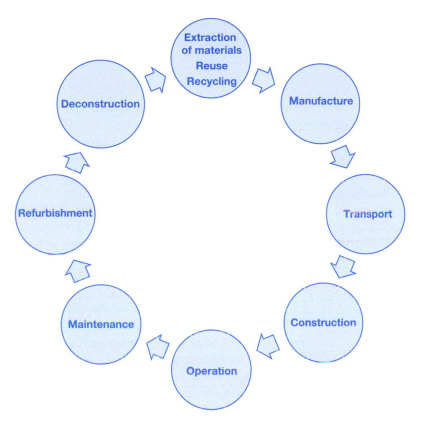

Figure 3.2 Lifecycle of a building from cradle to cradle.

3.3 What is embodied energy?

At its simplest, embodied energy can be defined as the total primary energy demand in all the lifecycle stages of a building, excluding operational energy. As discussed, measuring the lifecycle energy demand of a building includes that used in 'winning' (i.e. mining, extracting, transporting) raw materials, as well as the processing and manufacture of all the materials and components of a building during its construction, maintenance, refurbishment and disposal. Energy has two components: direct and indirect. Direct energy includes energy demand during construction and operation, such as fuels in machinery, electricity production, heat production and transport. In contrast, indirect energy includes the energy demand in processing materials and manufacturing products such as window frames, including processing equipment. Embodied energy is typically expressed in units of MJ or kWh.

In addition to primary energy, energy can also include feedstock energy. Feedstock energy describes the use of fossil fuels as a raw material in the manufacture of some materials. For example, gas and oil are raw materials in the manufacture of plastics. Embodied energy studies typically only take primary energy as a parameter, but some studies also include both primary and feedstock energy (Dixit *et al.*, 2010). A further subtle point of difference in how embodied energy is defined is that some studies interpret primary energy to be all energy, both non-renewable fossil fuels and renewable energy, whereas others only consider embodied energy to be that derived from non-renewable resources (Dixit *et al.*, 2012).

3.3.1 Embodied carbon

Embodied carbon represents carbon dioxide emissions (expressed as $kgCO_2$ or $kgCO_2e$) emitted as a result of primary energy use at each stage in a building's lifetime.

Embodied carbon is usually quantified in units of kilograms ($kgCO_2$) or tonnes (tCO_2) of carbon dioxide or carbon dioxide equivalent. In fact, three different but related units of measure are used: carbon (C), carbon dioxide (CO_2) and carbon dioxide equivalent (CO_2e). Carbon (C) is the fraction of carbon in CO_2 and is found by dividing CO_2 by 12/44 (the respective atomic and molecular weights). CO_2e is a more complete measure of greenhouse gases (GHGs). It enables the calculation of all the different GHGs based on the amount of warming (termed global warming potential, GWP) that a given amount of a specific GHG may cause, using CO_2 as a reference. For example, CO_2, as the reference gas, has a GWP of 1 and methane has a GWP of 25. So, reducing one tonne of methane is the equivalent to 25 tCO_2 or, to follow scientific convention, 25 tCO_2e. Emissions are given in different metric units, tonnes (t), kilograms (kg) or grams (g).

Embodied energy and embodied carbon are not quite analogous. Embodied carbon can also include other sources of GHG emissions such as CO_2 from chemical processes. For example, the carbonation of lime in cement production, hydrofluorocarbons (HFC), blowing agents used in the production of insulation, or nitrous oxide from fertilizer manufacture. While energy is always an input, carbon can also be sequestered, or locked away. Biogenic materials (such as trees, hemp or wool) and the products made from them can be thought of as a carbon sink (this is discussed in Section 3.7). Carbon can also be emitted by materials at the end of their useful life when they are recycled or disposed of, for example, when plastics or timber are burned and their carbon component is released into the atmosphere, or biological materials (e.g. plant or animal fibres) are landfilled, resulting in emissions of methane.

3.4 What do embodied energy studies tell us?

Lifecycle energy in buildings has been studied since the early 1990s. Studies have considered housing (e.g. Adalberth, 1997) and, less frequently, commercial buildings such as offices (e.g. Cole and Kernan, 1996; Junnila *et al.*, 2006). These typically compared different energy standards (e.g. Winther and Hestnes, 1999; Thormark, 2002), construction methods (e.g. Buchanan and Honey, 1994; Monahan and Powell, 2011) and construction waste management (Craighill and Powell, 1996).

A recent literature review of 36 published studies revealed that the mean embodied energy from the construction of residential buildings was 4.1 GJ/m² (206 cases) and 10.5 GJ/m² for commercial office buildings (based on 25 cases) (Yung *et al.*, 2013). Commercial buildings typically have a larger embodied energy per unit area than residential buildings. This is because they are required to bear higher loads per unit area and thus require greater structural support, employing materials such as steel and reinforced concrete which, typically, have a high embodied energy (Ramesh *et al.*, 2010).

Reviews of published studies consistently find a large range of embodied energy values (BIS, 2010; Yung *et al.*, 2013). For example, Yung *et al.* (2013) identified embodied energy values ranging from 1.5 GJ/m^2 (Winther and Hestnes, 1999) to 17.5 GJ/m^2 (Fay *et al.*, 2000) for residential buildings, and an even greater range for commercial buildings, from 1.4 GJ/m^2 (Buchanan and Levine, 1999) to 19 GJ/m^2 (Treloar *et al.*, 2000). Figure 3.3 provides a range of embodied energy values from a selection of published studies for housing.

What is clear from these studies is the lack of consistency in results. Some variation is expected due to differences in building type (e.g. single dwelling or multiple-occupancy) and construction type (e.g. timber frame, steel frame, load-bearing masonry) associated with cultural norms (Sartori and Hestnes, 2007). Furthermore, studies from different countries or

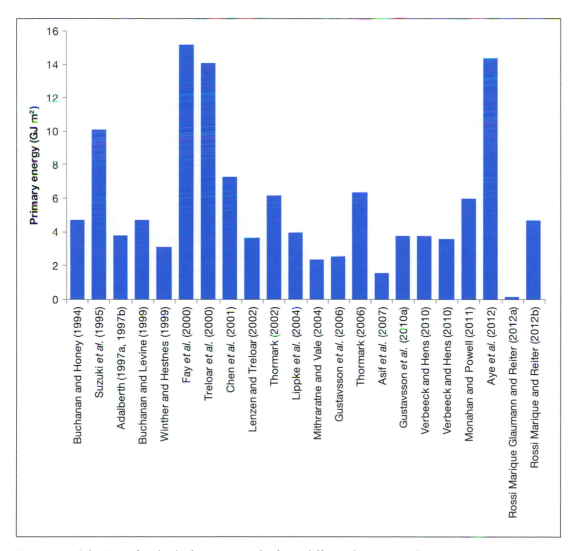

Figure 3.3 Selection of embodied energy results from different housing studies.

regions are not only based on different geographic and climatic characteristics, but also different industrial and economic production processes and energy generation fuel mixes, which add to the variation in embodied energy found. However, the wide variation in results can also be attributed to methodological inconsistencies relating to different assumptions between different studies, including the boundary and scope of the analysis and data (BIS, 2010; Sartori and Hestnes, 2007; Dixit *et al.*, 2010; Yung *et al.*, 2013). Indeed, the variation found is so great that each study could be said to be unique and, as such, incomparable.

In addition to the different lifecycle boundaries discussed, some of the variation can be attributed to differences in the scope, which relates to what is accounted for and what is excluded. Differences include: site workers' equipment and travel to and from the site; the energy consumed by ancillary services (e.g. site offices) and professional services (e.g. the architects); and the energy used in the production of the equipment and tools used. Other differences include whether the lifecycle assessment (LCA) is calculated in terms of primary energy or final energy demand, as well as the functional unit used. A functional unit defines precisely what is being studied and quantifies the service delivered by providing a reference to which the inputs and outputs can be related. The functional unit is important as it enables alternative goods or services to be compared and analysed on an equivalent basis. For instance, the functional unit for the study of a building can relate to the whole building, gross floor area, net internal floor area, heated area or internal volume.

There is also a lack of consistency in the assumed building lifetime (25, 30, 50 and 100 years are typical). This has consequences for the overall lifecycle energy results and the climate change impacts of the buildings, particularly if carbon sequestration is included (this is discussed in Section 3.7).

3.4.1 Data

Finding robust, good-quality and consistent primary data that can be used to calculate the embodied energy of buildings is difficult. Manufacturers of products used in construction do not, in general, calculate the embodied energy and carbon emissions for their products and, if they do, they seldom make this publically available. As a consequence, data on materials and products are often unavailable, leading to gaps in the analysis. Furthermore, due to the significant time resource implications of collecting the necessary primary data, researchers often have to resort to other data sources to fill these gaps. These include secondary data from generic databases, literature, proxy (or substitute) data or data derived from the researchers' or companies' own studies and embodied energy coefficients. In addition, data are often out-of-date or inconsistent with the study's requirements (e.g. different technological and economic contexts) (Dixit *et al.*, 2010).

Data resources are, however, becoming more widely available, particularly in Europe and the USA (Table 3.1), including publically available databases such as Bath University's Inventory of Energy and Carbon (Hammond and Jones, 2008), the European Reference Life Cycle Database (ELCD, 2010) and the US Life Cycle Inventory Database (NREL, 2012). In addition, generic industry-produced datasets are available, including steel (World Steel Association, 2011) and plastics (Plastics Europe, 2011). There is also a growing body of Environmental Product Declarations (EPDs) and published product LCAs for both specific and generic construction products that are consistent in how they are calculated (Moncaster and Song, 2012; CEN, 2012b).

Table 3.1 Selection of available databases for building-related materials and products

Database	Type	Geographical coverage	Access	Boundary	Lifecycle method	Developer
Athena	Construction	USA	Licensed	Cradle to gate	Process	Athena Institute
Building Products Life Cycle Inventory (BP LCI)	Construction	Australia	Open	Cradle to grave	Process	Building Products Innovation Council
Ecoinvent	General	International	Licensed	Cradle to gate	Process	Ecoinvent Centre
European Lifecycle Database (ELCD)	General	European	Open	Cradle to gate	Process	EU joint research
GaBi Lifecycle database	General	International	Licensed	Cradle to gate	Process	PE International
Bath University Inventory of Carbon and Energy (ICE)	Construction	UK	Open	Cradle to gate	Process	BSRIA
Embodied Energy and CO_2 Coefficients for NZ Building Materials	Construction	New Zealand	Open	Cradle to gate	Hybrid	Victoria University
US Life Cycle Inventory	General	USA	Open	Cradle to gate and cradle to grave	Input/ Output	National Renewable Energy Lab (NREL)
Wood for Good lifecycle database	Timber products	UK	Open	Cradle to grave	Process	Wood for Good

Source: Moncaster and Song (2012).

Comparing the results of different studies is difficult without clear information on where and when the data used were collected. The differences between studies are too great, and drawing direct comparisons or general conclusions from a comparison is, therefore, inappropriate (BIS, 2010). The results from such studies can only ever be indicative and should be interpreted with caution and careful attention to the methods used, the system boundaries applied and what has (or has not) been included before any interpretation can be made or conclusions drawn.

To improve comparability, many of the problems identified can be solved, in part at least, by a clearly defined standardised methodological approach to the study of the LCA of energy and carbon in buildings. A standardised method would facilitate a fairer comparison between different options, benchmarking of performance and, perhaps in the future, certification and regulation.

3.5 Measuring embodied energy and embodied carbon

LCA is the principal method used to calculate the environmental impacts of products, including complex assemblies of products like buildings, at all stages along the lifecycle of that product. LCA, which first developed during the late 1960s and early 1970s in separate methodological developments in the USA, UK, Germany and Sweden (Hunt *et al.*, 1996; Baumann and Tillman, 2004), is a relatively young discipline and is continually developing (Finnveden *et al.*, 2009). The first studies employed different methodologies, making comparisons difficult and raising questions concerning robustness. During the 1990s, a process of consensus building by the international LCA community culminated in the standardisation of the LCA methodology in International Standards to which LCAs should comply: ISO 14040:2006 to ISO 14044:2006 (Baumann and Tillman, 2004).

3.5.1 Lifecycle assessment methods

LCA and embodied energy studies use three different methods: process-based (bottom up); economic input–output (EIO; top down); or a hybrid of the two.

The process-based method accounts for all the environmental impacts, including direct energy inputs, of each contributing material and process in a linear sequence along the production lifecycle. However, process-based analyses have been criticised for excluding many other contributory processes from the wider economy, such as capital goods and infrastructure (Dixit *et al.*, 2010). Failure to capture these other inputs results in an underestimate of the true embodied energy and carbon.

The EIO method was developed to capture the processes from the wider economy, including the production of capital goods, the factories, roads and other infrastructure. EIO analyses use the flows of money between different sectors, in the form of input–output tables derived from national government data. Embodied energy and carbon are calculated by multiplying the cost of the product by the energy intensity of that product. This approach is assumed to give a more comprehensive and complete picture as it accounts for the entire system boundary. However, EIO LCA is limited. First, there will be uncertainty in how well a specific sector is modelled due to the aggregation of collections of industrial types into sectors. Second, EIO LCA models are often incomplete as, in general, they only use data that are publically available, which is typically limited to a small number of environmental impacts.

The third method, the hybrid analysis, uses the process-based method but expands the system boundary with the EIO method. This accounts for upstream processes where it becomes difficult to achieve reliable and consistent information, or where data are incomplete.

The process-based method is the most widely adopted LCA method in embodied energy studies of buildings (BIS, 2010). Furthermore, in the context of buildings, embodied energy methods are exclusively concerned with flows of energy and emissions of GHGs.

3.5.2 Whole-building lifecycle energy and carbon assessment standards

Despite LCA standards being formalised in International Standards (ISO 14044:2006), there has been and continues to be a lack of comparability of building lifecycle energy and carbon assessments due to lack of methodological standardisation. One possible explanation is that the ISO 14040:2006 and ISO 14044:2006 standards are generic for all products and services and do not prescribe specific methods. However, buildings are complex assemblies of materials and products with multiple lifecycles, and each whole-building lifecycle energy and carbon assessment may be subtly different (i.e. boundaries, scope, etc.) yet still comply with the standards. As a consequence, this lack of prescription has become an important driver for a standardised methodological approach. Currently (2015), there is no internationally accepted standardised method of accounting for the embodied energy and carbon of buildings. Nevertheless, the field is developing fast, with a number of voluntary standards being developed that are either generic standards or specific to buildings. These include the UK's Publically Available Specification (PAS) 2050 for the calculation of the carbon footprint of products (BSI PAS, 2011) and the European Committee for Standardisation Technical Committee 350 (TC/CEN 350) recently developed voluntary standards explicitly for buildings (EN 15978; CEN, 2012b) and construction products (EN 15804; CEN, 2012a).

The PAS 2050 (2011) specification for the assessment of the lifecycle GHG emissions of goods and services, developed by the Carbon Trust and DEFRA, is a specification for quantifying the carbon footprint of individual products across their supply chains 'from raw materials through all stages of production distribution, use and disposal/recycling' (BSI PAS, 2011, p. 2). This common specification used in assessments supports both comparability of different studies and also the compatibility of results in additive product chains. Thus, results from different products and processes can be added together along supply chains. Though not specifically for whole-building assessments, the energy and carbon impacts of a building can be calculated as the sum of the impacts of the individual products and processes in the construction of that building.

Based on the existing ISO 14040 and ISO 14044 standardised LCA methodology, the European Standards EN 15804 and EN 15978 (CEN, 2012a; 2012b) prescribe a standardised calculation methodology to assess the energy and environmental impact of construction products and buildings, respectively.

The method uses a modular approach to calculate the lifecycle embodied energy in construction products and whole buildings (Figure 3.4). The process-based lifecycle method has four stages (A to D), three of which, with their component modules, are illustrated in Figure 3.4 (A–C), together with a fourth, Module D, which extends beyond the system boundary. Module D includes the benefits and loads beyond the building lifecycle, such as the reuse, recovery and recycling of building materials and components into new products and components, which may in turn be used in the future substitution of resources. Inclusion of Module D would provide a cradle-to-cradle LCA of the building.

Modules A to C include environmental impacts directly linked to processes and operations within the system boundary of the building, comprising:

- Modules A1–A3: cradle-to-gate processes of materials and products used in construction.
- Modules A4–A5: construction processes from transport and storage of materials and products from factory gate to site, up to completion of the construction.
- Modules B1–B7: operational phase from completion of construction to end-of-life decommissioning.
- Modules C1–C4: end-of-life deconstruction and disposal processes.

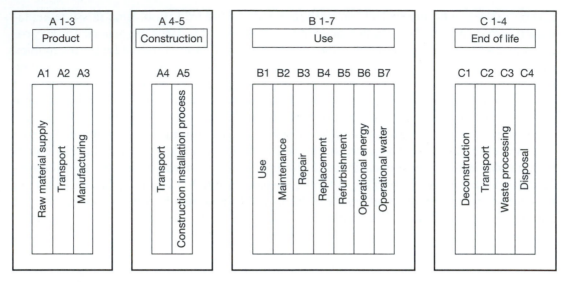

Figure 3.4 Definition of lifecycle stages of a building.
Source: CEN (2012b).

However, the standards are based on current industry practice, and current levels of energy efficiency and carbon intensity. They do not allow for future changes that may occur. For example, EN 15978 specifies an end-of-life based on current practice, assuming today's average landfill and recycling rates are sustained. But any new building currently assessed using this method is unlikely to reach this stage for at least 60 years, if not longer. Given current policy projections, landfill is likely to be an obsolete method of waste management by this point in time.

3.6 Reducing embodied energy

Measuring embodied energy in a building enables the identification of hotspots and can also suggest opportunities for reducing embodied energy and carbon. Such opportunities for reducing embodied energy can include:

- design geometry of building;
- material specification;
- logistically efficient transport or local sourcing;
- operation: construction operation efficiency; lower carbon construction operations;
- waste management.

The design of a building is perhaps the most significant influence on its embodied energy. The compactness of a building's form and size has significant implications for the volumes of materials used. A small building will almost inevitably have lower embodied energy than a large one as it will use fewer materials, walls and services. Fewer materials equates to lower embodied energy. Design decisions also influence how a building can be constructed, which in turn has consequences for both the energy embodied in actual construction and also the specified materials.

The specification of materials can also offer opportunities for reducing embodied energy. For example:

- Specifying materials with lower embodied energy, such as selecting a design using timber frame rather than steel or concrete frame.
- Specifying durable and long-lived finishes and technologies can reduce the need for maintenance and replacement.
- Specifying a material sourced from a supplier or manufacturer that has either conformed to a certification scheme (e.g. Forest Stewardship Council; FSC) or has voluntarily reduced energy or decarbonised extraction/manufacturing processes along their products' supply chains.

Secondary materials can also be specified, but this can be problematic as buildings and their components are generally not designed to be taken apart and reused at end-of-life, which limits the availability of useful materials. Increasing recycling rates and the use of secondary materials requires changes to how buildings are designed and how materials are recirculated. Much of a building at the end-of-life is unusable due to the construction and design detailing, which makes it difficult to separate components and materials. To enable increased availability of useful secondary materials in the supply chain, buildings and their components need to be designed for reuse and recycling, in conjunction with an established supply chain to improve sourcing.

Transport and construction operations are other areas that offer opportunities for reducing embodied energy. Transport mileage, though relatively insignificant, can be further reduced by efficient logistics or local sourcing of materials. The efficiency of construction can also be improved by increasing the number of components or even whole buildings to be manufactured off-site rather than conventional construction on-site.

The two connected areas that offer the greatest potential for reducing embodied energy are materials and construction methods, which we discuss in the following sections.

3.6.1 Embodied energy in materials

In general, the amount of embodied energy of a material is related to the amount of processing a material undergoes. Different materials require different amounts of processing to transform them from raw materials to products. Table 3.2 shows the embodied energy and embodied carbon values for typical construction materials. Highly processed materials (e.g. polyurethane insulation, steel, glass, cement) have relatively high embodied energy values per kilogram, compared to materials with minimal processing (e.g. timber). For example, flax-based insulation has higher embodied energy and carbon than sheep's wool insulation due to the significantly higher processing and longer transport chain of the flax insulation (Murphy *et al.*, 2008; Table 3.2).

Material substitution, in this instance changing from a high embodied energy material to one with lower embodied energy, can significantly reduce the embodied energy of a building and, if widely implemented, have significant impacts on global energy and carbon emissions (Buchanan and Levine, 1999). For example, when investigating Indian load-bearing

Table 3.2 Embodied energy (MJ) and carbon dioxide ($kgCO_2e$) of common construction materials by mass (UK averages) (2008)

Material		Embodied energy (MJ/kg)	Embodied carbon dioxide ($kgCO_2e/kg$)
Aggregates		0.10	0.00
Bricks (common)		3.00	0.24
Concrete blocks		3.50	0.24–0.38
Cement		4.50	0.74
Concrete (general)		0.80	0.11
Glass		15.00	0.91
Insulation	Fibreglass	28.00	1.35
	Polyurethane rigid foam	101.50	4.26
	Recycled sheep's wool	9.00	0.21
	Cellulose (recycled newspaper)	0.94–3.33	0.63
Plasterboard		6.80	0.39
Steel (general – average recycled content)		20.10	1.46
Timber (sawn softwood)		7.40	0.20–0.39

Source: Hammond and Jones (2008).

masonry buildings, Ventakarama Reddy and Jagdish (2003) found that substituting conventional brickwork with an unfired earth (soil–cement) block reduced the embodied energy of the building by up to 45 per cent.

In a typical UK residential building, concrete, brick and timber account for the majority (by mass) of materials used and contribute the majority of embodied energy and carbon emissions attributed to that building (Monahan and Powell, 2011). Concrete has been found to contribute 61 per cent of the embodied energy of a conventional masonry house (Asif *et al.*, 2007) and 36 per cent of a timber-framed house (Monahan and Powell, 2011), whereas in commercial buildings concrete and steel are the predominant materials (Xing *et al.*, 2008).

Three studies that compare the embodied energy (cradle-to-construction) of three different structural materials in residential buildings (Buchanan and Honey, 1994; Buchanan and Levine, 1999; Lippke *et al.*, 2004) clearly show that construction using higher embodied energy materials, such as steel or concrete, results in a building with significantly higher embodied energy when compared to timber in an equivalent building (Figure 3.5). Timber-framed construction consistently requires less embodied energy and emits less CO_2 to the atmosphere than concrete-framed construction (Gustavsson *et al.*, 2006; Lenzen and Treloar, 2002; Monahan and Powell, 2011; see the case studies in Section 3.8).

Material substitution is not always technically feasible. In such cases there may be opportunities to make changes to the production processes or the constituent materials that can reduce the embodied energy and carbon content. For example, the production of Ordinary Portland Cement (OPC), the 'glue' that holds concrete together, has a high

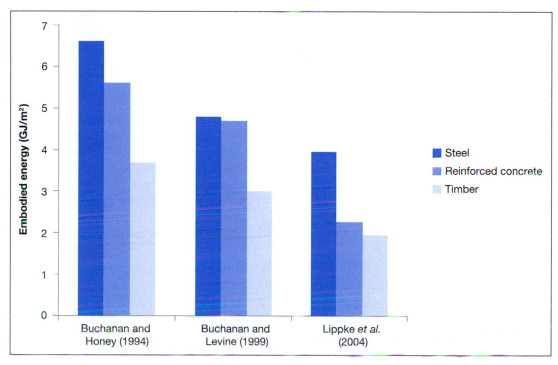

Figure 3.5 Comparison of embodied energy of steel, reinforced concrete and timber-frame construction (GJ/m²).

Source: Buchanan and Honey (1994); Buchanan and Levine (1999); Lippke *et al.* (2004).

embodied energy and contributes the majority of emissions associated with concrete. Taken as a global average, one tonne of cement has an embodied carbon of 830 kgCO$_2$ per tonne, with 40 per cent of these emissions derived from the mainly fossil fuel energy used in its production (IEA, 2009). The chemical process used in the production of cement (known as 'decarbonation') produces the remaining 60 per cent of GHG emissions and is unavoidable. The global average carbon figure differs from that shown in Table 3.2, highlighting the problematic nature of selecting data, as discussed in Section 3.4. The value in Table 3.2 is specific to the UK and reflects the lower carbon factor of production from differing fuel mixes and technologies used in production processes.

It has also been suggested that up to 98 per cent of fossil fuels used in the manufacture of OPC can be substituted by low-carbon alternatives, which can reduce emissions by 24 per cent globally (MPA, 2010). Further reductions in the embodied energy of concrete can be made by supplementing a proportion of the cement used in concrete with a lower embodied energy substitute, without compromising the technical characteristics required. Cement substitutes are secondary materials, usually derived from waste streams of other processes, and have lower embodied energy and carbon emissions than cement. They include:

- silica fume: a waste product from production of silicon metal and ferrosilicon alloys;
- pulverised fly ash (PFA): waste from coal-fired power production;
- ground granulated blast furnace slag (GGBS): from the production of iron and steel.

A recent example of the environmental benefits that substitution can achieve is the construction of the 2012 London Olympic Park, where OPC was substituted with lower-carbon alternatives (with PFA preferred over GGBS). Overall, there was a 32 per cent cement substitution, which resulted in an estimated 12 per cent reduction in embodied carbon (Henson, 2011).

Expanding the lifecycle boundary to capture the whole lifetime embodied energy of a building can radically change the embodied energy outcomes. For example, Aye *et al.* (2012) compared the whole lifecycle embodied energy of modular prefabricated steel multi-residential buildings with a concrete construction. The prefabricated steel building required around 50 per cent more embodied energy than the concrete building in the construction stage. However, expanding the study boundary to include the whole lifecycle, which encompassed end-of-life management, found structural steel offered significant reuse and recycling potential post-deconstruction compared with the equivalent concrete building. This resulted in an 81 per cent lower whole-life embodied energy compared with the concrete scenario.

A number of studies have assessed the potential energy saving of construction using secondary materials in buildings. Thormark (2000), when comparing a single-family house constructed with a large proportion of secondary materials and components, with an alternative house constructed using new materials, found an embodied energy saving through the use of secondary material of 40 per cent. Currently, the reuse of materials and building components, or even whole buildings, is not a routine practice in most countries. If embodied energy becomes a widely adopted concern, then designing for reuse can significantly reduce the resources demanded and environmental impacts of our built environment.

3.6.2 Construction systems

Changing how buildings are constructed can provide significant opportunities for reducing embodied carbon. Typically, buildings are constructed in-situ from materials that are brought in and 'cut to fit' on-site. An alternative approach is prefabrication, which is the production of building components, sections or whole buildings off-site for erection on-site. Increasing the use of prefabrication, or off-site manufacturing, of building components, or even whole buildings, creates a regulated production process that is more efficient and which may radically reduce the embodied energy of buildings.

First, for on-site construction, sheet materials and timber are cut to fit at the site itself, resulting in wastage. Furthermore, some materials arriving on-site may never be used due to being damaged or over-ordered, which also results in wastage. Indeed, in the UK, up to 13 per cent of construction waste is thought to be composed of clean, unused materials (Constructing Excellence, 2008). Research has suggested that substitution of traditional methods with prefabrication systems can reduce concrete and timber waste by 50–60 per cent and 74–84 per cent, respectively, resulting in significant reductions in the amount of materials required and, as a consequence, reduce embodied energy (Tam *et al.*, 2005; Jaillon *et al.*, 2009).

Second, off-site production may be more energy efficient than traditional on-site construction. In a comparison of conventional on-site construction with a modular off-site construction of timber-framed housing in the USA, off-site production reduced energy demand by 30 per cent (Quale *et al.*, 2012).

Third, off-site production processes may also involve changes in transport patterns associated with construction, resulting in fewer movements of larger loads of both materials and workforce, which may also reduce embodied energy. Quale *et al.* (2012) found 35 per cent fewer transport miles (including both material and equipment and workforce movements), equating to an overall 20 per cent reduction in transport-related carbon emissions compared with on-site construction.

Finally, current on-site practice brings a fluid workforce together for a specific construction project, which does not foster an environment for institutional learning and innovation. In contrast, off-site manufacturing requires a stable workforce. This stability can create a working environment for institutional learning and innovation (Miles and King, 2013).

In summary, compared to on-site construction, the potential benefits of off-site factory production include reduced material inputs, reduced waste, lower energy demand, fewer transport movements and improved quality control (Quale *et al.*, 2012; Miles and King, 2013).

3.7 Natural materials: sequestration and the consequence of time

Natural materials, such as timber, can be considered to be a special case in the context of embodied energy, and in particular embodied carbon, due to their often (although not always) low embodied energy and their ability to sequester carbon.

Before we consider sequestered carbon, we need to have an understanding of how carbon emissions relate to time and climate change. By convention, when accounting for carbon emissions it is assumed that one tonne of CO_2 emitted today has the same value (or climate change impact), whether it happens at the start or end of a building's lifetime and regardless of the given reference period (20, 50 or 100 years). However, it is the effect of cumulative carbon emissions over time that drives climate change (Stocker *et al.*, 2013). The longer a period of time that one tonne of CO_2 is in the atmosphere, the greater the climate change impact it causes. So delaying, or avoiding altogether, the emissions of one tonne of carbon for a given period will have a positive outcome compared to emitting that tonne of carbon today. One tonne of CO_2 emitted today will have a greater climate change impact over a given time period than if it was emitted 20, 50 or 100 years in the future. As a consequence, in order to take the effect of this cumulative impact into consideration, there needs to be a shift in the weighting of CO_2 in an LCA towards the early initial stages.

This is relevant to the way lifecycle energy in buildings can be accounted for. The majority of a building's embodied energy occurs upfront, prior to occupation (stages A1–A4, Figure 3.4). If a cumulative approach is used, each unit of carbon emissions arising from the construction of a building will have a significantly greater climate change impact than an equivalent unit of carbon emitted during the later stages of the building's lifetime. A cumulative approach will, therefore, result in an increase in the weighting of initial embodied carbon emissions because these will be more damaging over a given period of time. Weighting in this way is, however, generally not undertaken.

This argument is central to the consideration of natural materials and the special case of sequestration of carbon in bio-based products and materials in buildings. Natural, also referred to as bio-based materials, are renewable materials (e.g. wood, straw, hemp, wool)

Box 3.1 Worked example: timber

Timber, a principle construction material, has the following vital statistics:

- Density 500 kg/m^3
- Energy, approximately 17 MJ/kg
- Carbon content of approximately 50 per cent, depending upon species.

Therefore, 1 kg of dry construction-ready timber will contain:

1 kg × 0.5 = 0.5 kg carbon

This value can be converted to CO_2e by dividing by the ratio of the molecular weight of carbon to carbon dioxide (CO_2e) 12/44, equivalent to multiplying by 3.67:

0.5 × 3.67 = −1.85 kg CO_2e/kg

The result is shown as a negative as it represents a net removal of CO_2 from the atmosphere. However, this will only be the case if the timber is sourced sustainably and the net amount of forestry remains stable or increases. If its harvesting resulted in a net deforestation, then no net CO_2 removal will have occurred.

that take up atmospheric CO_2 during growth, locking it up, or sequestering it, as carbon material within their biomass (termed 'biogenic' carbon). This sequestered carbon may act as temporary carbon storage when it is incorporated into products or things such as buildings. In accounting for embodied carbon, it has been argued that sequestered carbon can be thought of as being a carbon reduction or 'negative emission' (Brandão *et al.*, 2013). Thus, in simple terms, a carbon footprint accounting for the net carbon arising from a product or building is the total embodied carbon minus the carbon sequestered and temporarily stored in a biogenic material.

The temporary storage of carbon sequestered in biomaterials is generally not taken into account and is excluded from carbon footprint and LCA studies of buildings. When using these methodologies, carbon is perceived as a flow to and from nature (i.e. the ground or atmosphere), with CO_2 uptake during growth assumed to equal CO_2 released at end-of-life. Hence, carbon neutrality is assumed and only emissions arising from the use of fossil fuels are usually accounted for.

When a cumulative emissions approach is taken, the identification of benefits to delaying emissions by temporarily 'storing' carbon in buildings can be included. As CO_2 is not released in today's atmosphere but at a future date (conventionally 100 years), it can be argued that some credit should be given for this delay in releasing the stored carbon. But this relies on assumptions concerning what has and will happen in the future.

First, there is the assumption that production is sustainable, that is to say the total resource is constant or is increased. For example, wood may come from clear felling virgin forest or it may be derived from sustainably managed forests. Furthermore, the production of many plant-based materials may involve use of fertilisers, pesticides and other processes (including growing, harvesting, transporting, processing and end-of-life decomposition) that may have substantial embodied energy and emissions.

Second, there are assumptions concerning the length of a building's lifetime and its final disposal. At the end-of-life of a building, a material may, for example, be combusted and the carbon it contains released into the atmosphere; or it may be landfilled, whereby a proportion of the carbon will remain buried but some will be converted to (and then released as) carbon dioxide and methane (roughly 50:50). Alternatively, the material may be reused, recycled or remanufactured, in which case the carbon continues to be sequestered.

So, should sequestered carbon count, and if so, how can it be counted? The issue is complex. As discussed, sequestered carbon and embodied carbon are different. Subtracting sequestered carbon from embodied carbon to give a single net carbon result may not be appropriate. Furthermore, presenting a single 'carbon footprint' figure conceals production impacts, making it difficult to account for individual stages or processes. To be meaningful and useful in benchmarking and target setting, any accounting method needs to be transparent, with the three elements (embodied, operational and sequestered) clearly stated.

Currently, the issue is subject to intense debate and there is no consensus on the inclusion of temporary carbon storage, let alone a standardised methodology for accounting for temporary carbon storage in buildings. The issue of how to take temporary carbon storage into account when quantifying the energy and carbon embodied in buildings highlights how important clarity and transparency are when accounting and reporting on embodied energy and carbon. Reporting a net carbon balance should be avoided and both embodied and biogenic/sequestered results should be reported separately.

3.8 Case studies

3.8.1 Case study 1: Lingwood, UK, off-site construction of low-energy housing (Monahan and Powell, 2011)

The first case study is a development of 15 low-energy, low-cost homes in Lingwood, Norfolk, UK. This case study illustrates the use of carbon footprinting as a tool to benchmark and evaluate the embodied carbon associated with the design of a housing development of low-energy housing in the UK.

Project background

- Location: Lingwood, Norfolk, UK
- Building type: residential
- Tenure: rental or shared ownership (affordable social housing)
- Year of construction: 2008
- Size: terraced with 71 and 83 m^2 internal floor area
- Construction approach: timber-framed structural insulated panels (SiPs) fabricated off-site.

The design was constrained by the client brief, which specified that the construction was to be flexible, affordable and buildable on a larger scale in different locations. The design criteria included the following:

- Highly insulated panellised system, supplemented with additional insulation installed on-site to exceed contemporary minimum building regulation standards.
- Sustainable material use including off-site manufactured timber frame, combining the benefits of sustainable materials and resource efficiency in production, together with timber (all certified FSC) cladding and window frames, which reduced the use of high embodied energy materials such as plastics, masonry and concrete.

Approach

The objective of the study was to assess the embodied energy and carbon of design decisions to inform future developments of affordable housing. The outcomes of the study were to:

- benchmark the embodied energy and carbon of the construction phase;
- compare the design as built with different aesthetic and construction approaches;
- share the knowledge gained to assist in decisions for future developments.

An LCA framework was used to benchmark the development and to compare it with two alternative design choices and construction practices. In total, three scenarios were used:

1 Off-site larch facade: constructed using off-site panellised system with larch facade.
2 Off-site brick facade: larch facade material substituted by brick.
3 Conventional masonry cavity: conventionally constructed house using masonry cavity construction.

For the analysis of embodied energy and carbon, a three bedroom semi-detached house design of 83 m² internal floor area was used. The boundary of the study was cradle-to-construction, encompassing emissions from the following: materials and products used in construction; final transport of the materials and products to site; waste produced on-site; transportation of waste to disposal; fossil fuel energy used on-site during construction; and fossil fuel energy used in components' manufacturing. The study did not include internal elements (e.g. internal walls, doors), finishes (e.g. paints, plasterboard, skirting board) and fittings (e.g. bathrooms, lighting, kitchens). It was assumed that these would be identical for all of the compared construction types and was therefore excluded from the analysis.

Findings

The study found that a house built using a panellised timber frame, constructed off-site, had 26 per cent lower embodied energy and 34 per cent lower embodied carbon compared with an equivalent traditional masonry house (Table 3.3). This was mainly attributed to

Table 3.3 Embodied energy (MJ) and embodied carbon (kgCO₂e) for off-site larch cladding, off-site brick cladding and conventional masonry cavity scenarios (2008)

	Total primary energy consumption (GJ/m² floor area)	Embodied carbon (kg CO₂e/m² floor area)
Scenario 1: off-site larch cladding	5.7	405
Scenario 2: off-site brick cladding	7.7	535
Scenario 3: conventional masonry cavity	8.2	612

Source: Monahan and Powell (2011).

the use of materials, in this case a higher proportion of timber, which has a relatively low embodied energy and low mass. It therefore requires less substructure than conventional buildings (Figure 3.6). Although the mass of the substructure of the Scenario 1 house was lower, half of the embodied carbon relating to materials was associated with the construction of the substructure, foundations and ground floor (Figure 3.7). This was because the relative importance of these substructural components increases with the reduced use of carbon-intensive materials, such as timber, in other components. For example, in Scenario 3, the conventional masonry cavity building, the proportion of embodied carbon attributed to substructural components is less than 35 per cent.

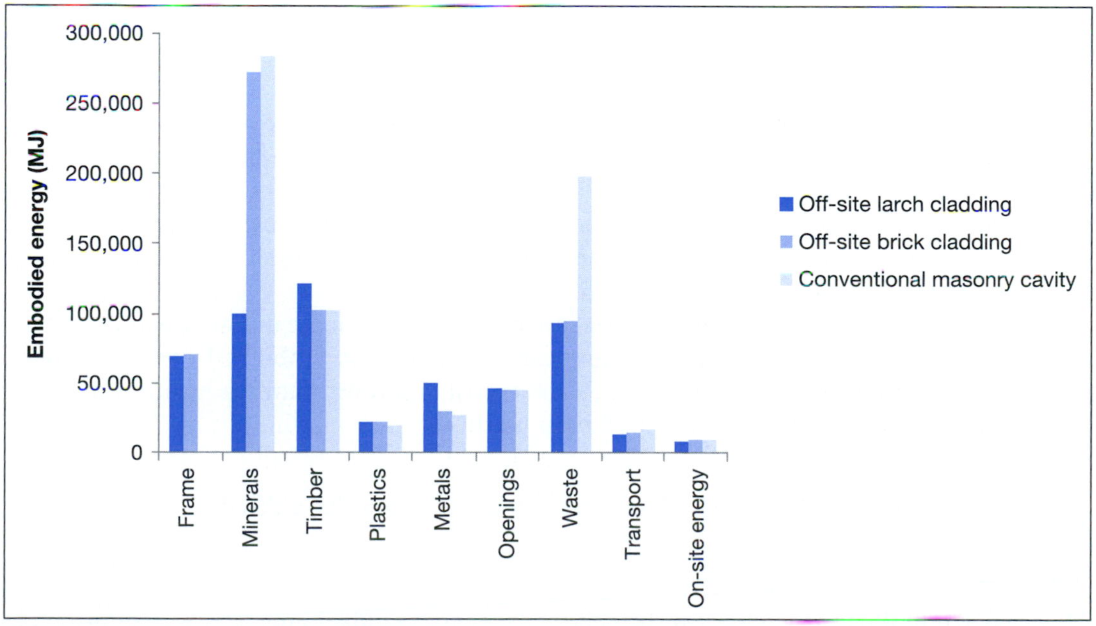

Figure 3.6 Embodied energy of different materials for off-site larch cladding, off-site brick cladding and conventional masonry cavity scenarios (MJ) (2008).

Source: Monahan and Powell (2011).

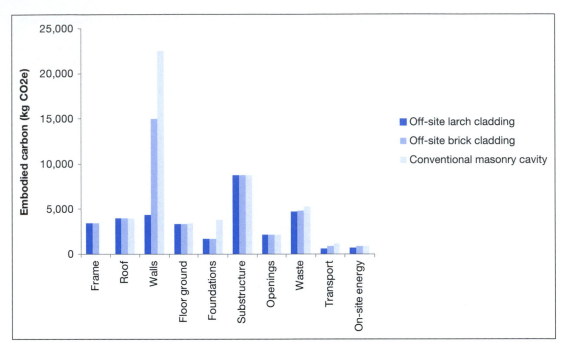

Figure 3.7 Embodied carbon by construction element for off-site larch cladding, off-site brick cladding and conventional masonry cavity scenarios (kgCO$_2$e) (2008).

Source: Monahan and Powell (2011).

Key messages

- The case study demonstrates the effects of material substitution on embodied carbon, with the reduction in embodied carbon relating primarily to the use of a lightweight timber frame.
- A secondary factor contributing to the reduction in embodied carbon was the use of off-site manufacturing, with its improved efficiency of volume production, and a reduction in the energy used and waste.
- By taking embodied energy and carbon into consideration in the initial design decisions, the construction project achieved significant reductions in embodied carbon without compromising other building criteria, such as cost and time, than if a business-as-usual conventional masonry approach had been used.

3.8.2 *Case study 2: MAKAR Ltd, Inverness, Scotland (Monahan, 2014)*

Rationale

This case study illustrates what can be achieved when embodied energy is a core principle in the construction of houses. In this example, sustainable low-impact materials, efficient production methods and localisation of supply chains are combined to reduce embodied energy.

Project background

- Location: Inverness, Scotland
- Building type: residential
- Tenure: single bespoke commissioned individual private homes and multi-house projects for social housing sector (rented)
- Year of construction: 2014
- Size: 20 homes per year in 2014
- Construction approach: off-site timber frame, structural panel system.

Since 2002 MAKAR Ltd have developed an approach which combines the efficient use of resources offered by off-site prefabrication with the use of sustainable low-impact materials and a commitment to developing local supply chains and the Highland economy. The selection of all materials, components and services was made with a full consideration of the following criteria, where possible:

- natural renewable materials that avoid complex and energy-intensive manufacturing processes and which have low embodied energy during production and use;
- low toxicity to reduce off-gassing;
- durability over an extended period for low maintenance and long life;
- cost and affordability;
- local materials used to reduce transport miles and advantage local economies;
- recycled materials.

The purpose of the embodied carbon study was to support an understanding of the carbon impacts of combining natural sustainable materials, local procurement and resource-efficient off-site construction, to evidence the environmental benefits and to identify potential improvements.

Approach

The goal of the study was to undertake an embodied energy and carbon assessment to benchmark the environmental impact of a MAKAR home and to compare the low-impact design and specification against the industry norm (as represented by the Lingwood case study illustrated above).

In order to benchmark the environmental impact of the MAKAR system, a case study was undertaken on four semi-detached homes at a site in Blairninich, Fodderty near Strathpeffer for the Highlands Small Communities Housing Trust, constructed during 2014. The homes were three-bedroomed semi-detached, with a floor area of 86 m^2.

MAKAR collated a comprehensive inventory of materials, products, energy, transport and waste required and produced in the production of the Fodderty development. The inventory was compiled from actual quantities measured directly during the off-site manufacturing process and the on-site construction of the case study homes. This included the foundations, floor, wall and roof panels, windows and doors. For comparative purposes the plumbing pipework, electrical system, internal fittings and finishes were excluded.

In addition to quantifying the embodied energy and carbon, an evaluation of the sequestered carbon was also made to calculate the overall carbon balance cradle-to-construction.

Findings

The total embodied carbon for a MAKAR Fodderty home was estimated to be 24.8 tCO_2e, approximately 289 $kgCO_2e$ per m^2. This is very low relative to other constructions of this type, and was, for example, 30 per cent lower than that reported in case study 1 in Lingwood.

The transportation of materials from manufacture to the MAKAR production facility was 6 per cent of total embodied carbon (18 $kgCO_2e$ per m^2) (Figure 3.8). This was higher than that typically found in other studies, which suggests an average of 3 per cent of total embodied carbon arising from transport. Recognising that construction choices can have extensive and potentially positive effects on the local economy of remote regions like that of the Scottish Highlands, careful attention was given to the sourcing of construction approaches, materials and components. Many of the products, however, are not currently manufactured in the UK and were sourced from mainland Europe; the cellulose fibre insulation was, for example, imported from the Netherlands. A further factor is the relatively remote UK location of the MAKAR facility, which increased the distances products and materials had to be transported.

Only an estimated one tonne of waste material was produced during construction, amounting to an estimated 4 $kgCO_2e$ per m^2. This was extremely low, and was virtually a zero-waste construction process. Waste arising from the typical construction of homes in the UK has been estimated to be approximately 19 per cent of total embodied carbon in construction, roughly equating to 76 $kgCO_2e$ per m^2 (Hammond and Jones, 2008). The majority of waste material was timber (48 per cent). This was burned at the point of

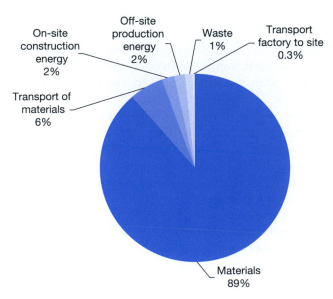

Figure 3.8
Embodied carbon in a MAKAR home by material, energy, waste and transport ($kgCO_2e$) (2014).

Source: Monahan (2014).

generation to provide space heating at the MAKAR facility (e.g. offices) or donated to staff members, family and friends for burning in log burners. Of the remaining waste, 34 per cent was recycled and 19 per cent of total waste exported as waste. The final disposal route of this material was not known, but was likely to have undergone separation, with the recyclable materials (e.g. plastics) reclaimed.

Timber was the predominant material used in this project, being used extensively in structural, fibreboard and insulation products (Figure 3.9). The use of timber from sustainably managed local Highland forests had a far-ranging positive influence on the carbon balance. The construction method employed made full use of locally grown timber that was certified, sourced, processed and fabricated locally into the following:

- structural components: joists, beams and posts;
- external finishes: cladding, soffit, fascia, decking;
- internal finishes: flooring; staircases, sills, skirtings and other internal fittings and finishes.

Avoiding highly processed materials, such as plastics, and specifying low-impact sustainable products, i.e. cellulose and sheep's wool insulation, were significant factors in reducing embodied carbon. In contrast, typical construction specifies insulation materials derived from high embodied energy materials such as plastics (e.g. polystyrene or phenolic foam insulation).

As the larch cladding was air dried rather than kiln dried, this also avoided energy and emissions. With a very high timber and wood-based products content, the homes at Fodderty sequestered more carbon at construction than that embodied in their construction.

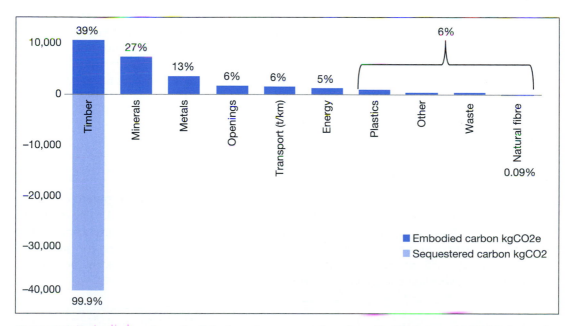

Figure 3.9 Embodied carbon (kgCO$_2$e) and sequestered carbon (kgCO$_2$) in a MAKAR home by material, energy, transport and waste (2014).

Source: Monahan (2014).

However, until the full lifecycle of the homes is known and what happens to these materials after deconstruction is established, the findings can only be indicative.

MAKAR homes typically have a type of foundation termed post and pad, which radically reduces the amount of concrete required, which is, of course, a high mass, high embodied energy material. In comparison, the typical approach taken in constructing timber-framed homes uses a conventional oversite concrete slab and footings (i.e. trenches following the footprint of the walls filled with concrete), which have a higher embodied carbon.

With all new homes being required to achieve ever increasing environmental and sustainability standards, this study showed that MAKAR's approach achieved a 30 per cent lower embodied carbon compared with conventional timber-frame off-site construction, providing an exemplar of one way of achieving this goal.

MAKAR have adopted many of the recommendations made to reduce embodied carbon, such as reducing the use of materials with high embodied energy, specifying low embodied carbon renewable materials that also act as carbon sinks, sourcing timber from sustainable sources, building components off-site, procuring materials and services locally, and reducing transport movements. This was reflected in the low embodied carbon of MAKAR at Fodderty and a near-zero waste construction.

Key messages

- This case study shows that combining off-site construction with low-impact materials, together with careful detailing and use of high embodied carbon materials (e.g. cement) and local supply chains, radically reduces embodied carbon without compromising quality.
- Specifying materials, such as timber and wood-based fibre products (e.g. cellulose fibre insulation and wood fibreboards), can sequester substantial quantities of carbon.
- The lack of locally produced products that meet high environmental standards resulted in an increase in the transport footprint compared with that typically found in other studies, suggesting a trade-off between environmental costs and benefits is sometimes necessary.

3.8.3 Case study 3: International Towers, Sydney, Australia

Rationale

Case study 3, Bangaroo South, Sydney, Australia is a brownfield redevelopment currently under construction (completion due 2017) (Jones Lang LaSalle, 2013). The project's objective is to achieve a 20 per cent reduction in embodied carbon emissions compared to standard construction practice by using LCA to identify opportunities for new and innovative design and construction approaches. This case study illustrates the benefits of benchmarking embodied energy and carbon in construction.

Project background

- Location: Bangaroo, Sydney, Australia
- Building type: mixed-use low-rise/high-rise development

- Tenure: commercial (retail, hospitality, offices) and residential
- Year of construction: under construction (completion 2017)
- Size: 490,000 m² floor area in total on a 7.8 ha site
- Client: Lend Lease
- Architects: Lord Richard Rogers and Ivan Harbour of Rogers, Stirk, Harbour + Partners.

The redevelopment and regeneration of the 7.8 ha area will be Australia's first large-scale carbon-neutral development.

Approach

By using a carbon footprinting methodology from an early stage, and involving all actors in the process, the expected outcomes include:

- reduction in the carbon intensity of the development in the construction phase;
- increased transparency and understanding of the impacts of day-to-day design and material procurement decisions;
- establishment of a legacy within the construction supply chains and broader industry, through the support of low embodied carbon products and design solutions;
- sharing the knowledge gained in reducing the construction sector's contribution to climate change and providing a replicable model for future developments to save embodied carbon through smarter design and material choices.

To achieve these outcomes the project's carbon reduction initiatives are divided into three areas:

1   materials and procurement (including manufacturing and delivery);
2   on-site construction and construction innovation;
3   planning and design.

The procurement tender process incorporated carbon-neutral aspirations by including an evaluation of the embodied carbon of the main materials using questionnaires and guidelines. The supplier guidelines provided information on ways to reduce the embodied carbon of products and materials. The guidelines also established a hierarchy focusing on the areas of dematerialisation (using less or no materials to provide the same function), material substitution, recycled content, material sourcing and transport. The questionnaires provided a way for tenderers and suppliers to suggest various initiatives that can be applied to the project. They enable ideas and data to be captured and for products to be assessed based on their potential embodied carbon savings.

The embodied carbon assessment, which was designed to be benchmarked against standard construction practice, measures and tracks the development's embodied carbon footprint. An LCA of each building element, component and the top 20 materials used on-site was undertaken. The assessment included estimating the cradle-to-grave embodied carbon of building elements, materials and components, plus the total lifetime operational energy to estimate the lifetime carbon footprint of the development. The process of data collection was considered key to facilitating the comparison and consideration of alternative strategies for selection and sourcing of materials, supplier initiatives, material optimisation, transport efficiencies and design initiatives in order to meet the overall 20 per cent embodied carbon reduction target.

At the design stage a bill of quantities and specifications did not exist. Proxy data, to act as a benchmark against which comparisons could be made, were sourced from two comparable developments and used to derive benchmarks for foundations, envelope and fit-out, and applied to area schedules for the development. Other benchmark data include construction site energy, waste, delivery of materials and final disposal, end-of-life emissions.

Findings

A preliminary benchmark carbon assessment estimated the lifetime embodied carbon of the Bangaroo South development to be 440,308 tCO_2e total (1.3 tCO_2e per m^2) (Figure 3.10), the largest contributors being the envelope (40 per cent) and maintenance (30 per cent).

The headline results included:

- 99 per cent reduction in on-site construction waste;
- 20 per cent reduction in carbon intensity of reinforcing steel used (approximately 4 per cent of the overall 30 per cent carbon reduction target);
- 65 per cent cement replacement within specific building elements using Bangaroo South as a test bed for emerging concrete technology.

Additional legacy benefits are also beginning to emerge. Suppliers have adopted market solutions that have been piloted at Bangaroo South, while other suppliers have produced embodied carbon assessments of their own products based on what they have learned through the Bangaroo Project.

The initiatives identified include dematerialisation, material substitution, logistics and design. It was established that most of the embodied carbon reduction initiatives through

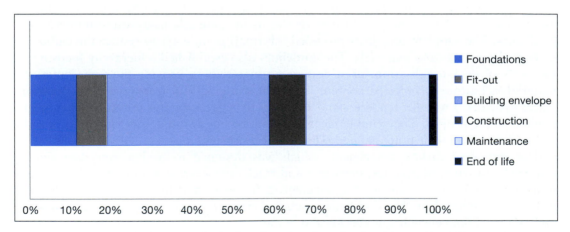

Figure 3.10 Estimated lifetime embodied carbon benchmarks for International Towers, Bangaroo South, Sydney Australia ($kgCO_2e$) (2013–2014).

Source: Deloitte LLC (2012).

the preliminary benchmarking can be achieved at low or no additional cost. Furthermore, some of the dematerialisation initiatives result in a cost saving. It is not yet known, however, what the total cost implications will be.

The embedding of the process of embodied carbon assessment into the development process assisted in the consideration of alternative strategies for the selection and sourcing of materials, supplier initiatives, transport efficiencies, material optimisation and design. The project also demonstrates the broader benefits and opportunities for innovation and learning that this approach can bring to the whole development process, not just design and construction, but also to investors, suppliers, contractors and clients.

Key messages

- By bringing the concept of embodied energy to the fore, from the initial tendering phase through to construction, in a process that includes innovation from all participants in the supply chain, cost-effective reductions in embodied energy and carbon have been achieved.
- It is too soon to know if the 20 per cent embodied carbon reduction objective will be achieved.
- This case study also demonstrates the additional benefits of undertaking an evaluation of the embodied carbon, with learning from the project providing a legacy to the supply chain beyond the lifetime of the project.

3.9 Summary

This chapter has shown that it takes a significant amount of energy to deliver a building. Yet the energy required to manufacture all of the materials, components and services required to construct and maintain a building are often overlooked, intentionally or otherwise. In addition, as buildings become more energy efficient and the operational energy demand reduces, the embodied energy becomes proportionally more significant. It also takes more energy to produce a low-energy building, more materials and, often, a greater number of technologies.

Quantifying embodied energy enables benchmarking and the identification of hotspots in the building that provide opportunities for reducing impacts. These opportunities include design, material specification and construction methods. However, boundaries, and what is included, are important to the outcome. An assessment that has too narrow a boundary may not provide a fair appraisal of the choices. A lifetime assessment from cradle-to-grave gives a more accurate assessment of the total embodied energy and environmental impacts of a design or material decision.

As a science, accounting for embodied energy in buildings is still in its infancy, but it is nevertheless developing quickly. Early studies were diverse in their methods and the data used, and their results variable and incomparable. Recent developments have seen improvements in the standardisation of methods, the increasing availability of consistently produced data and a new generation of readily available tools. These developments have significantly increased the ease of undertaking relatively complex whole-building assessments. In addition, the standardisation of methods and increasingly available data facilitate a fair comparison between different options and benchmarking of performance, and pave the way for certification and regulation of embodied energy in buildings.

In the future, the embodied energy of a building may be as well understood and have equal importance as operational energy does today.

References

Adalberth, K. (1997) Energy use during the life cycle of single-unit dwellings: Examples. *Building and Environment*, 32(4), 321–329.

Adalberth, K. (1997a) Energy use during the life cycle of buildings: a method. *Building and Environment* 32(4), 317–320.

Asif, M., Muneer, T. and Kelley, R. (2007) Life cycle assessment: A case study of a dwelling home in Scotland. *Building and Environment*, 42(3), 1391–1394.

Aye, L., Ngo, T., Crawford, R.H., Gammampila, R. and Mendis, P. (2012) Life cycle greenhouse gas emissions and energy analysis of prefabricated reusable building modules. *Energy and Buildings*, 47, 159–168.

Baumann, H. and Tillman, A.M. (2004) *The Hitch Hiker's Guide to LCA: An Orientation in Life Cycle Assessment Methodology and Application*, Studentlitteratur AB, Lund, Sweden.

BIS (2010) Estimating the amount of CO_2 emissions that the construction industry can influence: Supporting materials for the Low Carbon Construction IGT report. Department for Business, Innovation and Skills. www.gov.uk/government/uploads/system/uploads/attachment_data/file/31737/10-1316-estimating-co2-emissions-supporting-low-carbon-igt-report.pdf, accessed 9 June 2015.

Blengini, G.A. and Di Carlo, T. (2010) The changing role of life cycle phases, subsystems and materials in the LCA of low energy buildings. *Energy and Buildings*, 42, 869–880.

Brandão, M., Levasseur, A., Kirschbaum, M.F., Weidema, B., Cowie, A., Jørgensen, S. and Chomkhamsri, K. (2013) Key issues and options in accounting for carbon sequestration and temporary storage in life cycle assessment and carbon footprinting. *The International Journal of Life Cycle Assessment*, 18(1), 230–240.

BSI PAS (2011) PAS 2050: Specification for the assessment of the life cycle greenhouse gas emissions of goods and services. British Standards Institute, London.

Buchanan, A.H. and Honey, B.G. (1994) Energy and carbon dioxide implications of building construction. *Energy and Buildings*, 20(3), 205–217.

Buchanan, A.H. and Levine, S.B. (1999) Wood-based building materials and atmospheric carbon emissions. *Environmental Science and Policy*, 2(6), 427–437.

CEN (2012a) EN 15804: Sustainability of construction works. Environmental product declarations – core rules for the product category of construction products. BSI Group, London.

CEN (2012b) EN 15978: Sustainability of construction works. Assessment of environmental performance – calculation method. BSI Group, London.

Chen, T.Y., J. Brunett, C.K. Chau (2001) Analysis of embodied energy use in the residential building of Hong Kong. Energy, 26, 323–340.

Cole, R.J. and Kernan, P.C. (1996) Life-cycle energy use in office buildings. *Building and Environment*, 31(4), 307–317.

Constructing Excellence (2008) Construction and sustainable development. In Plain English guide to sustainable construction. http://constructingexcellence.org.uk/wp-content/uploads/2015/02/SUSTAIN GUIDE.pdf, accessed 27 May 2015.

Craighill, A.L. and Powell, J.C. (1996) Lifecycle assessment and economic evaluation of recycling: A case study. *Resources, Conservation and Recycling*, 17, 75–96.

Deloitte LLP (2012) *Deloitte D-Carbon8*, Deloitte LLP, London.

Dixit, M.K., Fernández-Solís, J.L., Lavy, S. and Culp, C.H. (2012) Need for an embodied energy measurement protocol for buildings: A review paper. *Renewable and Sustainable Energy Reviews*, 16(6), 3730–3743.

Dixit, M.K., Fernández-Solís, J.L., Lavy, S. and Culp, C.H. (2010) Identification of parameters for embodied energy measurement: A literature review. *Energy and Buildings*, 42(8), 1238–1247.

ELCD (2010) European Reference Life Cycle Database 2.0. http://eplca.jrc.ec.europa.eu/?page_id=126, accessed 9 June 2015.

EU (2008) Directive 2008/98/EC of the European Parliament and of the Council of 19 November 2008 on waste and repealing certain Directives. *Official Journal of the European Union*, 22 November.

Fay, R., Treloar, G. and Iyer-Raniga, U. (2000) Life-cycle energy analysis of buildings: A case study. *Building Research and Information*, 28(1), 31–41.

Feist, W. (1997) A comparison of life cycle balances: Low-energy house, Passive House, energy-autarchic house. In Research Group for Cost-efficient Passive Houses Protocol, *Material Selection, Ecology and Indoor Air Hygiene*, Passive House Institute, Darmstad.

Finnveden, G., Hauschild, M.Z., Ekvall, T., Guinee, J., Heijungs, R., Hellweg, S., Koehler, A., Pennington, D. and Suh, S. (2009) Recent developments in life cycle assessment. *Journal of Environmental Management*, 91(1), 1–21.

Gustavsson, L., Pingoud, K. and Sathre, R. (2006) Carbon dioxide balance of wood substitution: Comparing concrete- and wood-framed buildings. *Mitigation and Adaptation Strategies for Global Change*, 11(3), 667–691.

Hammond, G. and Jones, C. (2008) Embodied energy and carbon in construction materials. *Proceedings of the Institution of Civil Engineers: Energy*, 161, 87–98.

Henson, K. (2011) The procurement and use of sustainable concrete on the Olympic Park Learning legacy. Olympic Delivery Authority. http://learninglegacy.independent.gov.uk/documents/pdfs/procurement-and-supply-chain-management/01-concrete-pscm.pdf, accessed 27 May 2015.

Huberman, N. and Pearlmutter, D. (2008) A life-cycle energy analysis of building materials in the Negev desert. *Energy and Buildings*, 40(5), 837–848.

Hunt, R.G., Franklin, W.E. and Hunt, R.G. (1996) LCA: How it came about. *The International Journal of Life Cycle Assessment*, 1(1), 4–7.

IEA (2009) *Cement Technology Roadmap 2009: Carbon Emissions Reductions up to 2050*. IEA, Paris.

Jaillon, L., Poon, C. and Chiang, Y. (2009) Quantifying the waste reduction potential of using prefabrication in building construction in Hong Kong. *Waste Management*, 29(1), 309–320.

Jones Lang LaSalle (2013) International Towers, Sydney. Jones Lang LaSalle, Sydney.

Junnila, S., Horvath, A. and Guggemos, A.A. (2006) Life-cycle assessment of office buildings in Europe and the United States. *Journal of Infrastructure Systems*, 12(1), 10–17.

Lenzen, M. and Treloar, G. (2002) Embodied energy in buildings: Wood versus concrete – reply to Börjesson and Gustavsson. *Energy Policy*, 30(3), 249–255.

Lippke, B., Wilson, J., Perez-Garcia, J., Bowyer, J. and Meil, J. (2004) CORRIM: Life-cycle environmental performance of renewable building materials. *Forest Products Journal*, 54(6), 8–19.

Miles, J. and King, D. (2013) *Off-site Construction: Sustainability Characteristics*. Buildoff-site, London.

Mithraratne, N., B. Vale (2004) Life cycle analysis model for New Zealand houses. Building and Environment, 39, 483–492.

Monahan, J. (2014) MAKAR Carbon Measurement Project: An embodied carbon study of House Units 1 and 2 at Fodderty, Highland manufactured and erected for the Highlands Small Communities Housing Trust. A report prepared for MAKAR Ltd, UEA.

Monahan, J. and Powell, J.C. (2011) An embodied carbon and energy analysis of modern methods of construction in housing: A case study using a life cycle assessment framework. *Energy and Buildings*, 43(1), 179–188.

Moncaster, A.M. and Song, J.Y. (2012) A comparative review of existing data and methodologies for calculating embodied energy and carbon of buildings. *International Journal of Sustainable Building Technology and Urban Development*, 3(1), 26–36.

Mossman, C., Kohler, N. and Jumel, S. (2005) Life cycle analysis of Passive Houses. Paper presented at the 9th International Passive House Conference, Lidwigshafen-Darmstadt.

MPA (2010) A carbon reduction strategy. Mineral Products Association. http://cement.mineralproducts.org/current_issues/climate_change/greenhouse_gas_reduction_strategy.php, accessed 9 June 2015.

Murphy, R.J., Norton, A. and Campus, S.K. (2008) Life cycle assessments of natural fibre insulation materials. Imperial College London, prepared for the National Non-Food Crop Center (NNFCC).

NREL (2012) U.S. Life Cycle Inventory Database. National Renewable Energy Laboratory. www.lcacommons.gov/nrel/search, accessed 19 November 2012.

Plastics Europe (2011) Eco-profiles and Environmental Product Declarations of the European Plastics Manufacturers V2.0. www.plasticseurope.org/documents/document/20110421141821-plasticseurope_eco-profile_methodology_version2-0_2011-04.pdf, accessed 9 June 2015.

Quale, J., Eckelman, M.J., Williams, K.W., Sloditskie, G. and Zimmerman, J.B. (2012) Construction matters. *Journal of Industrial Ecology*, 16(2), 243–253.

Ramesh, T., Prakash, R. and Shukla, K.K. (2010) Life cycle energy analysis of buildings: An overview. *Energy and Buildings*, 42(10), 1592–1600.

Rossi, B., A.F. Marique, M. Glaumann, S. Reiter (2012) Life-cycle assessment of residential buildings in three different European locations, basic tool. *Building and Environment*, 51, 395–401.

Sartori, I. and Hestnes, A. G. (2007) Energy use in the life cycle of conventional and low-energy buildings: A review article. *Energy and Buildings*, 39(3), 249–257.

Stocker, T. F., D. Qin, G. K. Plattner, M. Tignor, S. K. Allen, J. Boschung, A. Nauels, Y. Xia, V. Bex, and P. M. Midgley (2013) *Climate Change 2013: The Physical Science Basis. Working Group I Contribution to the Fifth Assessment Report of the Intergovernmental Panel on Climate Change, Summary for Policymakers*, Cambridge University Press, Cambridge.

Suzuki, M., T. Oka, K. Okada (1995) The estimation of energy consumption and CO2 emission due to housing construction in Japan. *Energy and Buildings*, 22, 165–169.

Tam, C.M., Tam, V.W.Y., Chan, J.K.W. and Ng, W.C.Y. (2005) Use of prefabrication to minimize construction waste: A case study approach. *International Journal of Construction Management*, 5(1), 91–101.

Thormark, C. (2002) A low energy building in a life cycle: Its embodied energy, energy need for operation and recycling potential. *Building and Environment*, 37(4), 429–435.

Thormark, C. (2000) Environmental analysis of a building with reused building materials. *International Journal of Low Energy & Sustainable Building*, 1.

Thormark, C. (2006) The effect of material choice on the total energy need and recycling potential of a building. *Building and Environment*, 41, 1019–1026.

Treloar, G., Fay, R., Love, P.E.D. and Iyer-Raniga, U. (2000) Analysing the life-cycle energy of an Australian residential building and its householders. *Building Research and Information*, 28(3), 184–195.

Venkatarama Reddy, B. and Jagadish, K. (2003) Embodied energy of common and alternative building materials and technologies. *Energy and Buildings*, 35(2), 129–137.

Verbeeck, G., H. Hens (2010a) Life cycle inventory of buildings: a calculation method. *Building and Environment*, 45, 1037–1041.

Verbeeck, G., H. Hens (2010b) Life cycle inventory of buildings: a contribution analysis. *Building and Environment*, 45, 964–967.

Winther, B.N. and Hestnes, A.G. (1999) Solar versus green: The analysis of a Norwegian row house. *Solar Energy*, 66(6), 387–393.

World Steel Association (2011) Worldwide life cycle inventory for steel products. www.worldsteel.org/steel-by-topic/life-cycle-assessment/about-the-lci.html, accessed 9 June 2015.

Xing, S., Xu, Z. and Jun, G. (2008) Inventory analysis of LCA on steel- and concrete-construction office buildings. *Energy and Buildings*, 40(7), 1188–1193.

Yung, P., Lam, K.C. and Yu, C. (2013) An audit of life cycle energy analyses of buildings. *Habitat International*, 39, 43–54.

Zabalza-Bribián, I., Capilla, A.V. and Usón, A.A. (2011) Life cycle assessment of building materials: Comparative analysis of energy and environmental impacts and evaluation of the eco-efficiency improvement potential. *Building and Environment*, 46(5), 1133–1140.

4 Energy performance gap

4.1 Introduction

The decisions made concerning a building's form, orientation, fabric and services strongly influence its lifetime performance (Sorrell, 2003). The energy demand of buildings, together with the consequent emissions of carbon, are dependent on complex relationships and numerous interconnected actors, including government policies, energy utilities, house designers, planners, construction companies and occupancy. Steemers and Yun (2009) argue that building energy performance is primarily determined by interactions between three elements: building and climate characteristics; technological installations; and the occupants of buildings. Building characteristics include the type and amount of thermal insulation and the airtightness, while technological installations include the heating (and potentially electrical) technology employed, its efficiency, levels of control and the fuels used.

Overall improvements in energy efficiency have been achieved in many countries by tightening building standards, yet there often remains a significant difference between the predicted designed performance of a building and how it performs in reality. This difference is termed the energy performance gap.

The size of this gap varies considerably. In an assessment of 23 supposedly 'energy efficient' buildings, Bordass *et al.* (2004) found the actual energy demand to be twice that predicted at the design stage. More recently, the Carbon Trust (DECC, 2011) identified that the operational demand of buildings can be five times higher than the compliance calculations. A recent audit (UCL Energy Institute, 2013) of commercial buildings found the mean actual demand for office heating was 59 per cent higher than the design's mean estimated demand and 70 per cent higher for electricity demand (i.e. energy performance gaps of 59 per cent and 70 per cent). For educational buildings, the performance gap was 48 per cent for heating and 90 per cent for electricity.

In reality, it is quite difficult to identify the magnitude of an energy performance gap due to design teams not generally predicting actual in-use performance, as calculations are usually based on compliance to standards rather than performance (DECC, 2011). In addition, different calculation methodologies are likely to lead to dissimilar predicted levels of energy demand, while variations in actual energy use can be due to the alternative ways buildings and technologies are appropriated. We therefore suggest that although differences between expected and actual energy demand definitely exist, comparing averages in energy performance gaps (which inherently make the situation context-free) can be less useful.

Numerous influences underlie the performance gap, including those associated with commissioning, energy modelling and building construction, as well as occupation itself (e.g. hours of occupation, the rebound effect). The causes of the performance gap are therefore based not only on physical factors associated with the buildings themselves, including the quality of design and construction, the technical systems and local climate, but also on how the buildings are practically commissioned and used. It is not surprising that there can also be inaccuracies in the energy modelling of buildings. The difficulty of representing a complex system such as a building using a relative simple model is very likely to lead to uncertainty.

Several influences on the performance gap are also associated with poor communication and insufficient feedback, such as between: clients and designers at the briefing stage; designers and energy assessors; and construction companies and sub-contractors or occupants. For example, changes in how a building is constructed due to value engineering (i.e. delivering the same function at lower costs) can lead to modifications to the design which are not fed back to the energy assessors (ARUP, 2013). Also, if the information provided by a developer about how a building should be controlled is inadequate, the occupants or managers of the building may be unable to operate it efficiently.

To address these concerns the UK government, in collaboration with the Zero Carbon Hub, established the '2020 Ambition', which aims to 'demonstrate that at least 90% of new homes will meet or perform better than the designed energy/carbon performance' (Zero Carbon Hub, 2013, p. 3). This led to a work programme that will examine the construction of homes 'in its broadest sense, from conception through to completion on site' (Zero Carbon Hub, 2013, p. 10), including the gap between the energy demand a home is designed to achieve (the expected demand) and the actual energy it uses (the as-built performance). However, the work programme does not include the building occupation stage or non-residential buildings.

Based on the views of a wide range of experts, the Zero Carbon Hub (2013) examined the issues that contribute to the energy performance gap in residential buildings. This chapter is structured around the framework developed by the Zero Carbon Hub (2013), which is based on the stages of the construction process. However, we also consider the post-handover stage, exploring how occupants influence the energy demand of their buildings and non-residential buildings. There are, of course, considerable overlaps between the stages, and several factors influence more than one stage.

The following sections explore the energy performance gap at the following stages: concept and planning; design; materials and products; procurement; construction; testing and occupation.

4.2 Concept and planning

Decisions made at the concept and planning stage of construction influence the energy performance of a building. Problems can occur, for example, when the overall aim to produce a low-energy building is not clearly conveyed to the design team and onwards through the following stages. This is particularly the case when the stages involve different organisations or individuals (Carbon Trust, 2012).

One key element of the initial concept and planning phase is energy performance modelling, which is often separate from the design process. As a consequence it may not feed into the design choices. Decisions made at the concept stage can also create practical difficulties at the construction stage in terms of energy performance, but these issues are often not fed back to designers or modellers. One of the reasons for this is that professionals involved at the concept and planning stage may no longer be involved in the development project (Zero Carbon Hub, 2013).

4.3 Design

The traditional method of designing and constructing buildings is considered to be a linear process (client → architect → engineer → contractor → occupant), with each stage influencing the next (Cantin *et al.*, 2012) and where the consideration of energy efficiency does not necessarily play an important role. However, although the order of design and construction tasks is essentially linear, the various roles are highly interlinked, with numerous feedback loops. For example, the decisions of architects are influenced by how they think the building will be used by the occupant and the skills that they think the construction industry have to turn their design into a reality.

Another complication is that in many cases the design team is different from those involved in concept and planning. The aesthetics, fabric performance and building services are often considered separately, with this lack of integration extending to energy performance. This can lead to airtightness approaches being compromised and inadvertent thermal bridging (Zero Carbon Hub, 2014) (see Section 2.4.4 for thermal bridging and Section 2.4.5 for airtightness).

There is often a lack of knowledge concerning how decisions about design and aesthetic preferences affect energy performance and practical 'buildability' (Zero Carbon Hub, 2014). The design team may specify details that are difficult or impractical to build and have to be modified on site, which can result in a lower performance. In particular, details such as balconies can make airtightness more difficult, leading to compromises at the construction stage, especially when there is a lack of detailed design. Modifying designs, such as using standard sizes of materials to reduce excessive 'cutting to fit' can aid buildability, improve airtightness and reduce waste (Wingfield *et al.*, 2008).

It has been suggested that some designers, assessors or engineers of building services lack knowledge about energy performance in buildings (Zero Carbon Hub, 2013), but it is more likely that they only have knowledge of one or two approaches to an issue and lack the knowledge or experience of alternative approaches. This is especially important when we consider the multi-actor relationships at play here. For example, the architect may be skilled at one approach, such as designing airtight homes, but the construction firm may be skilled at another approach (e.g. Code for Sustainable Homes compliance and solar PV), meaning that one 'actor' may struggle to live up to the expectations of other actors.

During the early stages, designers often have to make significant assumptions and estimates concerning a building's energy demand due to having insufficient information about details of the building (Demanuele *et al.*, 2010). Changes may be made to the original specification later in the process due to, for example, changing requirements from the client or value engineering exercises (Bordass *et al.*, 2004). However, these variations, which can include changes to the thermal mass, insulation, orientation, controls and operational hours (ARUP, 2013), may not be reassessed in terms of energy performance.

Another reason for an energy performance gap is the misunderstanding that calculations for regulatory compliance with Building Regulations include all energy uses in the building, rather than just 'regulated' loads, which are limited to space heating and cooling, hot water, lighting, fans and pumps. This misunderstanding has been termed a gap in *perception*, rather than in *performance* (ARUP, 2013). Unregulated energy loads (e.g. server rooms, small power loads, lifts, catering and external lighting) may not be included at the design stage, yet can account for more than 30 per cent of the electricity demand in office buildings. For example, in a case study of a six-storey office building, the unregulated load amounted to 69 per cent of electricity demand (ARUP, 2013).

The majority of design software packages do not currently have energy demand calculations embedded in them. This process is, therefore, often undertaken by an energy assessor, leading to infrequent calculations. Smaller developers, who rely on an assessor to provide design and specification guidance, may receive variable standards of advice, depending on the assessor's expertise and ability. Insufficiently precise requirements for performance characteristics of materials and products may also lead to procurement teams purchasing lower-performing alternatives (Zero Carbon Hub, 2013).

What is required is a tool that is a better predictor of as-built performance than the current Standard Assessment Procedure (SAP), with particular emphasis on thermal mass, ventilation, cooling, lighting and hot water. Future simulation tools also need to model the impact of occupants and energy management on the energy performance (Menezes *et al.*, 2012). In addition to improvements to the model, there is also a need to develop further the integration of the tool into the concept, planning and design phases. There have been some improvements in modelling software (Menezes *et al.*, 2012) with management tools, such as Building Information Modelling (described in Section 4.9.1), but these have not been widely adopted.

It also needs to be remembered that these models are only as good as the input data. Considerable time and expertise is required to undertake an accurate simulation of a building (ARUP, 2013). The direction that models take us in is dictated by the decisions made by the modeller themselves. Modellers may also be constrained by the models due to only limited prescribed options being available within the model structure.

We also need to consider that while it is important to seek improvements to models, they will never be able to predict actual demand accurately, especially because usage can be contextual and dynamic. Technologies are not necessarily used in exactly the same way by everyone, so having a model input that can accurately predict energy demand will be difficult to attain. Inevitably, then, the way models are constructed, and thus problems defined (e.g. how to measure the performance gap), pushes us towards certain 'solutions' on how to solve that problem (e.g. how to narrow or even get rid of the performance gap).

4.4 Materials and products

Basic materials (e.g. bricks, cement) and products (e.g. window frames, heat pumps) used in construction have a significant effect on the energy performance of buildings. Although the performance of these products has to meet International Standards, they are usually tested as individual components in laboratory or factory conditions, and not as constructed on-site, where they may perform differently (Zero Carbon Hub, 2013). This is particularly the case when U-values are calculated rather than measured, which is what normally happens. Although this is acceptable under current regulations, more complex materials together with the demand for better thermal performance requires more sophisticated testing (ARUP, 2012). The on-site performance of materials and products is not generally considered in modelling software.

Another issue is a lack of guidance from the manufacturers on the installation and commissioning of products, as well as a lack of knowledge regarding the installation of innovative materials and products (Gupta and Dantsiou, 2013). It can also be difficult to distinguish between similar-looking products with different performance levels once packaging has been removed on-site, leading to incorrect usage and thus performance (Zero Carbon Hub, 2013). It has also been found that products intended to be used in one location were often used in another, leading to the need for modifications to ensure a reasonable fit, but which may result in reduced airtightness (Wingfield *et al.*, 2008) or poorer insulation.

4.5 Procurement

The complexity of the procurement process varies with the size of the developer and the development, the type of contract and the expertise and expectations of all those involved in the project. Although procurement teams base their purchases on a range of criteria in addition to cost, many have limited expertise in energy performance or are not provided with sufficient information concerning the performance levels required (Zero Carbon Hub, 2014). To reduce costs, materials of a lower specification may, therefore, inadvertently (or perhaps even intentionally) be purchased; there is no guarantee that this decision will then be fed back to the energy assessors or the design team. This problem can be compounded when purchases are made by sub-contractors rather than through the main contractor.

However, the provision of more information, as suggested by the Zero Carbon Hub (2014), may not necessarily lead to improved performance. It has been suggested (Foulds, 2013) that professionals are likely to acquire knowledge on energy performance (e.g. for procurement) through experience and learning by doing, rather than through being exposed to lots of factual energy performance knowledge.

Poor communication between different actors regarding the required level of energy performance and clarifying who is responsible for airtightness and pressure testing results can lead to building service contractors compromising the airtightness of a building. This can be a particular problem in Passivhaus buildings (see Chapter 6). There can also be issues relating to the lack of feedback to the design team and SAP assessor on changes to specified materials (Zero Carbon Hub, 2013).

4.6 Construction

The energy performance of a new or retrofitted building can be quickly undermined by small defects, such as gaps in insulation. In the past, small defects have largely gone unnoticed or have been ignored. However, increased concern with fuel bills and carbon emissions has led to the need for better quality construction (Tofield, 2012), but the construction team may lack energy performance related knowledge and training. This can lead to a culture that is inconsistent with achieving high standards of construction and energy performance. Inexperience, especially when linked to an imprecise design, can easily lead to greater fabric heat loss and thermal bridging than intended.

In addition to a lack of knowledge, there may also be a lack of willingness (conscious or not) to meet the energy performance levels required. For example, some builders do not appear to take airtightness requirements seriously, thinking the way they have always done it is acceptable, even though they may have the skills to construct airtight buildings. To address this the contractor must want to deliver, and insist on the workforce achieving, high-quality construction rather than there being the need to introduce a major new on-site initiative (Tofield, 2012). On-site quality control measures can, however, vary and there can be a lack of clarity as to who is responsible. If quality control is delegated to sub-contractors, the requirement for continuous overall monitoring can be lost. For example, the need to pass an air pressure test can result in the use of retrospective sealant fixes, rather than airtightness being achieved through good prospective building design and high-quality construction (Menezes *et al.*, 2012). Quality control has also been made more complicated by moving to a low-energy context. For example, is it now one person's job to ensure energy performance is met, or is it a shared burden/responsibility in everyone's job roles? Many of these responsibilities did not exist before and were consequently not considered.

Some build quality issues, such as improving the building fabric through better construction techniques, are gradually being tackled by the construction industry (Menezes *et al.*, 2012). Nevertheless, there is still concern with the installation of relatively new systems (e.g. heat pumps, mechanical ventilation systems) that contractors may not have sufficient knowledge and expertise to install and commission the services to the required standard (Zero Carbon Hub, 2013).

A barrier to increasing knowledge and skills is the tendency for construction companies to have only a small permanent workforce, resulting in frequent changes in temporary labour personnel. This has implications for the suggestion made earlier that individuals are more likely to learn through experience, rather than by being provided with information. In the long term, a stable workforce will be advantageous if individuals gain experience and exposure to these new 'low energy' ways of constructing. A workforce in flux is definitely a challenge when trying to meet the high energy performance expectations of a construction project.

The relationship between these complex temporary partnerships of sub-contractors is important and can lead to issues of trust. It is often considered necessary for contracts between these organisations to be detailed and complex, but this can lead to adversarial relationships (Winch, 2000), especially if there is the threat of litigation, which can arise if low profit margins lead to the desire to cut costs. Equipment may be over-specified to take this into account, so as to ensure there are high safety margins in performance (Sorrell, 2003). For example, radiators were installed in the bedrooms of a low-energy housing development because of a concern that the ground floor underfloor heating might be insufficient, despite energy performance calculations indicating the contrary. Perhaps unsurprisingly, these radiators were never actually used post-installation (Monahan, 2013).

One approach that can improve the relationships between the different actors and ensure the contractor has a longer-term involvement with the building is the 'Soft Landings' initiative developed by Building Services Research and Information Association (BSRIA) (see Section 4.9.6).

4.7 Testing

Testing is required to measure the airtightness of a building and to identify construction and product failings that will lead to reduced performance and thus contribute to the energy performance gap. The Zero Carbon Hub (2013) identified the following testing methods, but in the UK only air pressure testing is mandatory.

- Air: airtightness pressure testing; indoor air quality tests; ventilation flow rates.
- Thermal: thermal imaging; U-value measurements; co-heating test; hotbox.
- Building services: components; systems; controls.

It should be noted that a co-heating test measures heat loss from an unoccupied building that is heated to a reference temperature (typically 25 °C). One of the problems with this test is that it needs to be undertaken during the winter months, when there is a sufficient difference in internal and external temperatures. The hotbox measures the transfer of heat through building elements such as windows, doors and walls. There are only two facilities for hotbox testing in the UK (ARUP, 2012).

The uncertainties of these tests are often not clearly stated and in some instances there is a lack of standard practices. Rather than these individual methods, which test only one aspect of performance, the development of one overall energy performance test is preferred. Although testing can result in improved energy efficiency of individual buildings, its real value lies in the feedback to designers and construction teams, with the aim of improving future design and construction practices and thus reducing the energy performance gap (Menezes *et al.*, 2012). This, however, rarely occurs.

4.8 Occupation

The energy demand of a building is dependent not only on its characteristics and equipment, but also on how the occupants actually use the building and its technologies. Case studies have demonstrated that considerable variation exists between individual households (e.g. Monahan and Powell, 2011), including Gram-Hanssen (2004) finding that energy demand of similar households can vary by up to 300 per cent.

The occupants (and managers) of a building have a significant influence on energy demand as they regulate, for instance, the internal temperature, lighting, hot water, ventilation and equipment usage. Thus, numerous aspects of occupation can influence the performance gap. In considering this further, we now explore patterns of occupancy, operation of control systems, energy efficiency and the rebound effect, en-route to exploring an alternative approach to occupancy and the energy performance gap.

4.8.1 Patterns of occupancy

There is a high level of uncertainty about the assumptions made in energy usage predictions concerning patterns of occupancy. For example, the hours of use, heating schedules and number of occupants can have a significant influence on energy demand. In a study of 15 schools, a wide range of key factors influencing energy use was found, including occupancy patterns, internal temperatures, equipment electricity load and usage (Demanuele *et al.*, 2010). For example, it might be assumed that a school is occupied only during the school teaching day (from 0830 to 1530 in the UK), but in reality it may be used 'out-of-hours' by teachers, school clubs and the local community. Design assumptions may also overlook community use during school vacations. In the schools study, the occupancy schedule across the schools ranged from 31 to 51 hours/week, the heating schedule 30 to 79 hours/week and the office equipment schedule 7 to 24 hours/day (Demanuele *et al.*, 2010). In the same way, the 24/7 global economy may mean that office buildings are occupied late into the evening rather than just the assumed 'office hours'.

4.8.2 *Operation of control systems*

Occupants and building managers can find their building difficult to understand and control, leading to higher energy demand than anticipated (Demanuele *et al.*, 2010; Bordass *et al.*, 2004). The occupants of all buildings need clear information about how to operate their building and its installed systems, especially the control systems, effectively. This is even more important for buildings that are different from the norm, such as Passivhaus buildings, and especially large buildings that have more complex control systems. It is essential that a control system is designed with the competency of the operator(s) in mind. An engineer may install a control system that is straightforward to them, but beyond the knowledge of the likely operator. A complex control system may undermine an occupant's confidence and lead to a lack of willingness to engage with understanding the system, let alone maximising the energy efficiency of the building.

A higher level of energy demand may be due to the control system not working correctly, or it may be the case that the system needs fine-tuning, as shown in the Elizabeth Fry case study (Section 4.10.1). New occupants may lack the knowledge to optimise the controls of a large building and may not have been provided with sufficient technical knowledge by the contractor. This problem can be addressed by the use of a 'Soft Landings' approach, in which the contractor retains involvement in the building post-occupancy (see Section 4.9.6).

4.8.3 *Energy efficiency and the rebound effect*

As buildings become more thermally efficient, less energy is required for heating so occupants will, in theory, have lower heating bills and thus save money. But if instead of saving money, the occupants heat their buildings to a higher temperature than prior to, for example, a retrofit, the energy savings may be less than expected. This is termed the rebound effect. If the occupants were underheating their buildings prior to the improvements, then it is good that they are now able to afford to be warmer, especially if they were previously in fuel poverty. However, if an occupant is relying on energy savings to repay the capital cost of the thermal improvements, as in the case of the UK Green Deal, less savings are available to repay the loan.

The rebound effect was initially explored by Jevons in 1865, who pronounced 'it is wholly a confusion of ideas to suppose that the economical use of fuel is equivalent to diminished consumption. The very contrary is the truth' (Jevons, 1865, chapter 7.3). 'But such an improvement of the engine, when effected, will only accelerate anew the consumption of coal' (chapter 7.21). In the context of using coal to generate steam, he argued that as the process became more efficient it led initially to a reduction in coal demand, which in turn led to lower coal prices. This resulted in coal becoming economically viable for new uses, which ultimately greatly increased coal demand and enabled the economy to grow. Jevons (1865) showed that a decline in the real price of energy led to a powerful impetus for generating economic growth.

A fascinating study of the history of lighting (Fouquert and Pearson, 2006) also illustrates how increased efficiency leads to increased consumption. From the year 1300 to 2000, lighting efficiency evolved markedly from medieval candles to eighteenth-century oil lamps, gas lighting in the nineteenth century and electric lighting in the twentieth century. This led to a reduction in the cost per lumen (a measure of light) and thus an increase in the demand for light and therefore energy. More recently, the introduction of more efficient compact fluorescent lamps (CFLs) and now light emitting diodes (LEDs) has led to an explosion in lighting innovation (although not, so far, a big rise in energy use for lighting). We have also changed from one light fitting in the middle of a room to uplighters, downlighters, task lighting, security lighting, garden and patio lighting.

Increasing energy efficiency can be seen to reduce the cost of producing goods and services, leading to greater wealth, which in turn increases energy demand to higher levels than before the energy efficiency measures were introduced (Rosenow and Galvin, 2013). Energy efficiency improvements lead the consumer to use cost savings to purchase other goods and services, while the producer uses cost saving to increase output and thus increase the consumption of capital, labour and materials.

A rebound effect can be direct or indirect. Direct rebound involves paying for more of the same energy service, such as using more energy to warm your house further. If the financial savings are spent on different services, such as driving further or flying abroad, this is called indirect rebound, which also has energy use and carbon emission consequences. At a macro level, 'economy-wide' rebound can also be direct or indirect. For example, if a steel-making process is made more energy efficient, it might be assumed that it will lead to energy savings. However, a more efficient process will reduce the cost of steel production and thus it can be sold at a lower price. This might in turn lead to increased sales, an increase in the quantity of steel produced and therefore an increase in energy demand. This is direct rebound. Alternatively (or in addition), the low cost of producing steel may lead to cheaper cars and therefore to an increase in car travel (Sorrell, 2007). This is an example of indirect rebound.

Rosenow and Galvin (2013) identified three reasons for the energy savings from energy efficiency measures in homes being less than anticipated: the rebound effect; the pre-bound effect; and the technical quality of the retrofit, such as inadequate application of insulation material. Pre-bound occurs when the energy used in a home prior to retrofitting is lower than the calculated value; there is, therefore, less energy and money to save. A review of eight studies of 3,400 German homes (Sunikka-Blank and Galvin, 2012) found the measured energy used was on average 30 per cent lower than the calculated values; the lower the efficiency of the buildings, the larger the difference between the measured and calculated values. For example, a home with a calculated rating of 200 kWh/m^2 per year and a measured demand of 160 kWh/m^2 per year has a pre-bound of 20 per cent (Rosenow and Galvin, 2013).

A UK retrofit programme (TSB, 2013) found the average primary energy use of 37 social housing properties was 20 per cent higher than the national average. Post-occupancy monitoring showed substantial energy and carbon savings of 42 per cent and 49 per cent, respectively. However, these fell far short of the forecasted savings, which for CO_2 were 76 per cent. This was mainly due to the airtightness of the properties before the retrofit being better than expected, with most being below 10 m³/(h·m²) at 50 Pa, providing less opportunity for improvement than anticipated. However, 13 of the properties had an airtightness of more than 15 m³/(h·m²) at 50 Pa (TSB, 2013). To put this in context, the current regulatory requirement for new housing in the UK is a maximum of 10 m³/(h·m²) at 50 Pa, whereas the standard in Sweden is 1.5–3 m³/(h·m²) at 50 Pa (see Section 2.4.5 for airtightness).

Numerous studies have been undertaken to quantify the extent of rebound. In a review of over 500 studies, Sorrell (2007) found the effects varied between different technologies, sectors and income groups, with an average of less than 30 per cent direct rebound for heating across all households and 10 per cent for rich households. Other studies found a range of 12 per cent direct rebound for owner-occupiers, as well as a range of 49 per cent for low-income tenants in German homes (Madlener and Hauertmann, 2011). In the UK, rebound effects from insulation measures have been found to be 30 per cent for the average household (Milne and Boardman, 2000), and between 65 and 100 per cent for fuel-poor households (Hong *et al.*, 2006).

4.8.4 *An alternative approach to exploring the energy performance gap*

In this section we consider further why and how people use energy, because it is these influences that should guide us when we seek to lower operational energy demand, as part of minimising the energy performance gap. We explore alternative ways of thinking about this issue that contrast with the mainstream ways of defining and approaching occupational energy demand, critically reflecting on the principal approaches to addressing the performance gap. While this is mainly concerned with occupational demand, the messages are also relevant to other stages of the construction process.

The common and albeit often implicit assumption of research and policy is that energy use is a rational choice. We are said to use energy on the basis of a process of rational decision-making, which involves the weighing up of advantages and disadvantages of whether to act in particular ways. In the context of the energy performance gap, such an assumption will therefore lead us to thinking of the gap as something that not only can, but should be closed as it is not 'rational' or 'logical' to have paid for energy-saving technologies and then not to make the expected savings. It is perhaps unsurprising, then, that efforts to tackle the occupation-related influences of the performance gap have centred on information provision. The argument behind this is that to get the most out of low-energy technologies, people only need to realise that there is a net benefit in using those technologies in the 'correct' (energy saving) way.

Information provision commonly focuses on either (1) the benefits of saving energy (e.g. saving money, helping the environment, health benefits), or (2) the 'education' of occupants, perhaps through 'feedback' (as emphasised in Section 4.9.5) regarding how best to use their energy-demanding technologies; the assumption is that they want to save energy, but just do not know how to do so. In a similar vein, many initiatives are concerned with trying to make technologies easier to use; it is believed that 'getting the technologies right' will ensure occupants fall into line and close the performance gap by using the 'easy-to-use' technologies in the 'correct' way. Again, the assumption is that knowledge is the barrier and that there is an information deficit.

This line of thinking has dominated built environment policy for decades. However, do people make rational decisions all the time and always use technologies on the basis of information, cost and ease-of-use? By recommending an optimisation of technologies and education, the mainstream policy and research agenda assumes that this is the case. Similarly, most post-occupancy evaluations (POEs), but not necessarily all, are conducted on these assumptions. This is in spite of evidence that increasingly shows this is *not* the case (Guy and Shove, 2000; Guy, 2006). For instance, most people know that having the lights turned on uses energy and hence costs money, but yet does everyone turn the lights off when they leave the room? If we were to tell people that turning down their thermostat by one degree Celsius will save them 10 per cent on their heating bill, will they turn it down or is there something deeper in play regarding thermal comfort?

This mainstream approach implicitly ignores the rebound effect (Section 4.8.3), or assumes it can be overcome with better education and design. Alternative perspectives in the literature argue that the rebound effect is inevitable and that the energy performance gap will never be able to be reduced to nothing. One such perspective regards energy demand and the use of energy-saving technologies as being tied to ordinary, everyday activities (known as 'practices', such as cooking, showering, cleaning and hosting guests) that people do as part of their daily lives. People only demand energy and technologies so that they can perform practices. Thus, as low-energy technologies are unlikely to change the core rationale behind why someone wants to perform a practice, energy demand is unlikely to reduce by very much. Indeed, building occupants may take the opportunity (consciously or not) to achieve ways of living that they have always aspired to, but have not been able to achieve, such as higher indoor temperatures, or buying a new iPad. Following this argument, Gram-Hanssen *et al.* (2012, p. 71) argue for more of a focus on everyday life in the context of the rebound effect, recommending that we 'develop measures that have proven successful in real life on how to introduce new efficient technologies to users without resulting in practices changed to achieve higher norms and expectations and thus growing energy consumption'.

All this is not to say that we do not need energy-efficient technologies. The provision of well built, highly insulated homes will, of course, help to reduce energy demand. Rather, proponents of this alternative focus on everyday life argue that it can be used to re-frame our expectations of low-energy technologies and of the energy performance gap. Instead of focusing so intently on technological solutions, they argue that we need to understand more deeply the underlying influences that fundamentally underpin energy demand. For example, the energy performance gap exists not because individuals do not make the correct rational choices, but because of social conventions associated with comfort (e.g. thermal comfort in one's home), cleanliness (e.g. the need to shower and wash one's clothes regularly) and convenience (e.g. being able to use energy when and how one wants) (Shove, 2003).

4.9 Initiatives to reduce the energy performance gap

A variety of initiatives have been developed, although not necessarily widely adopted, to inform and help reduce the energy performance gap. Many of these focus on improving the flow of information between the various actors, and involve a change from the traditional, somewhat linear process, described above, to an integrated team approach to design and construction. Some initiatives concentrate on particular stages of the construction process, such as design or post-occupancy, while others have a more cross-cutting approach. We discuss the following initiatives that are most commonly advocated: building information modelling (BIM); construction team integration; benchmarking; commissioning; post-occupancy evaluation; and the Soft Landings approach.

4.9.1 *Building information modelling*

The need for integrated project teams and improved flows of information between team members has led to the development of IT-based tools such as BIM, which extend beyond the three-dimensional modelling of a building to include the modelling and management of information. This includes data concerning the properties of a building's components and its construction. The use of a building can also be included, enabling the efficient management and maintenance of the building during occupation.

BIM is intended to transform how teams and team members work together, making the database and how information is shared as important as the model. BIM facilitates the sharing of information not only between the project participants, but also between different stages of design, construction and operation, thereby enabling collaborative working. For example, an engineer is able to access information from the architect to prepare energy calculations (BIM Task Force, 2013). In addition to improving flows of information, the use of BIM should also reduce the energy performance gap by ensuring that changes in the products or materials used in a building are reflected in the energy performance models (this issue occurred in the Malmesbury Gardens case study, Section 4.10.2).

There are several different 'maturity levels' of BIM (RIBA, 2012):

- *Level 0* is essentially a paper-based process with computer-aided design (CAD) drawings.
- *Level 1* includes the use of a two- or three-dimensional CAD package with an IT-based approach to data structure and format. The cost/finance package that includes the commercial data is likely to be standalone, i.e. not integrated.
- *Level 2* requires the often fragmented team approach (see Section 4.9.2) to be replaced by teams working collaboratively under new forms of procurement. It requires the sharing of three-dimensional information models between key members of the team. However, these models do not have to be integrated into a single model. A standardised spreadsheet, called Construction to Operation Information Exchange (COBie), has been developed to gather and structure data (BIM Task Force, 2013).
- *Level 3* requires the combination of the various models into an integrated whole. This links together the design, cost, energy performance, asset management, and health and safety. It also enables potential iterations to the design to be easily assessed and compared in terms of cost and energy performance.

A UK government report found BIM to have 'the greatest potential to transform the habits and eventually the structure of the industry' (Innovation and Growth team, 2010,

p. 66) and estimated annual savings of £2bn to the industry and its clients (HM Government, 2012) through the widespread adoption of BIM. Subsequently, the UK BIM Task Group was created to support and help deliver the public sector's capability in BIM, with the aim that all publically procured projects will be delivered with the support of BIM level 2 by 2016.

4.9.2 Construction team integration

Building performance has been identified as being characterised by segmentation of professional and trade responsibilities, and fragmentation of activities such as preliminary and detail design, building performance simulation, the production of working drawings, tendering, planning and scheduling, commissioning and construction (Cantin *et al.*, 2012). The complexity of these relationships is considered to be a barrier to the construction of energy-efficient buildings (Cantin *et al.*, 2012) and to contribute to the energy performance gap, but the interactions between these different professions and trades are rarely considered.

The need for an integrated team to design and construct energy-efficient buildings is not a new concept. In 1998, the main recommendation of the Construction Task Force 'Egan Report' (Egan, 1998) was the integration of the construction process and team. Indeed, around that time it was considered difficult to find such teams: 'architects are rarely willing to compromise their design integrity and defer to M & E (mechanical and electrical equipment, or building services) requirements at an early stage' (1999 case study interview in Sorrell, 2003, p. 869). Four years after the Egan Report, the Strategic Forum for Construction (Egan, 2002) recommended that 20 per cent by value of UK construction projects should be undertaken by integrated teams and supply chains by the end of 2004 and 50 per cent by the end of 2007, but this does not seem to have occurred (Cabinet Office, 2012).

Although the Egan Report strongly encouraged industry to adopt a culture of performance measurement (Egan, 1998), and seven Key Performance Indicators were developed by the Task Force (Egan, 2002), the measurement of the thermal performance of the building was not included. However, more than two-thirds of projects identified in the Construction Best Practice Programme reported improved partnering, procurement or supply chain management skills in their organisation (Egan, 2002).

Not all studies, however, support the premise that fully integrated teams are essential for the effective delivery of projects, such as buildings that meet their energy performance targets. The organisational and behavioural barriers to full integration, such as traditional attitudes, the temporary nature and often changing composition of teams, can be considerable (Baiden, 2006). Teams managed by award-winning managers may, for example, work flexibly and collaboratively without achieving full integration (Baiden, 2006).

4.9.3 Benchmarking

If a building does not perform as well as predicted it is often the cause of some embarrassment to the design and construction teams. Not surprisingly, this undesired outcome is unlikely to be publicised, so lessons are unlikely to be learned. If this situation is to change there needs to be more openness and a willingness to learn from past mistakes.

To address this issue in the UK, the Royal Institute of British Architects (RIBA) and the Chartered Institution of Building Services Engineers (CIBSE) developed an energy benchmarking website, CarbonBuzz (www.carbonbuzz.org/index.jsp). This allows organisations to anonymously submit and compare the design and actual energy data for their individual projects with CIBSE benchmarks, thus building up a national database on building energy use. It is hoped the website will raise awareness of the performance gap and encourage users to develop more accurate estimates of operational energy use at the design stage.

There can be a problem with benchmarks as they can become out-of-date very quickly, but the intention is that data submitted to CarbonBuzz will be used to inform the benchmarks, keeping them up-to-date. However, when comparing energy data with benchmarks, what does it mean to be 'average' or 'poor' compared with the case studies? Further, an audit of CarbonBuzz data (UCL Energy Institute, 2013) found that the design energy data rarely include unregulated loads, which are considered to be one of the main reasons for the performance gap. However, the CarbonBuzz website now assists users to identify the unregulated energy used in their buildings.

4.9.4 Commissioning

In the context of the built environment, commissioning is concerned with making buildings work as they were intended to. Typically, commissioning refers to the setting up, testing and adjusting of both hardware (e.g. boilers, pumps, thermostats) and software (e.g. electronic control strategies), providing an opportunity to optimise the operational energy demand of mechanical systems. Commissioning is a systematic quality assurance process that ensures a new or existing building meets defined requirements through testing and verification.

Poorly commissioned systems may run inefficiently, resulting in excessive energy demand and inadequate performance, contributing significantly to the energy performance gap. The main advantage of effective commissioning is that it can reduce energy and operational costs. Indeed, it is often cited as the single most cost-effective strategy for reducing energy, costs and carbon emissions in buildings (Mills, 2011; Wang *et al.*, 2013).

Mills (2011), in a review of commissioning projects of 643 non-residential buildings in the USA, found effective commissioning reduced total primary energy demand by an average of 16 per cent. In addition, the cost of commissioning had a payback of 1.1 years, representing a significant return on the investment over the lifetime of the building.

In its narrowest sense, commissioning is defined as 'does this work as it was designed to?' This begins and ends with a predetermined system design and patterns of usage, and is explicitly concerned with ensuring systems work efficiently and safely. Yet commissioning can also have a much broader definition, taking a holistic systems approach that emphasises the connection between equipment and the 'softer' elements, such as the logic underlying controls, the effectiveness of user interfaces and energy trends within the building. This broader definition of commissioning differs from standard practice. Rather than focusing on working with pre-determined designs, it can also facilitate the asking of fundamental questions such as 'Is this needed?' and 'Can this be done differently?' at earlier design stages. Such early questioning opens the design process up from 'business-as-usual' optimising of conventional approaches and creates room for innovation.

While commissioning is becoming more prevalent, driven by its inclusion in Certification Schemes and Building Codes and Regulations (for example in the USA, commissioning is mandatory for all federal buildings and for Leadership in Energy and Environmental Design (LEED) certification), it still remains a relatively uncommon practice.

4.9.5 Post-occupancy evaluation

Post-occupancy evaluation (POE), also called building performance evaluation, assesses the performance of a building after it has been handed over to the occupants. The term is often interpreted rather differently and as a consequence it can be applied rather differently, depending on the objectives of the study and what exactly is being evaluated. Although several different approaches can be used, a POE is usually based on the collection and analysis of either (or sometimes both) quantitative and qualitative data. The latter often include occupant questionnaires as part of the evaluation of occupant experiences and opinions regarding how well the building meets their needs. Quantitative data include environmental measurements, such as air quality, temperature, noise, light, relative humidity and ventilation, and energy demand data, which are compared with the design energy performance. A POE can also identify ways in which the building design and performance can be improved. In addition to describing POE in terms of a general approach to project evaluation, there is also the more specific approach undertaken by the Post-Occupancy Evaluation consultancy (www.postoccupancyevaluation.com).

Three components of a POE have been identified: feedback, feedforward and benchmarking (Menezes *et al.*, 2012). Feedback measures the performance of a building and informs the occupants or managers of how they are performing, whether there are any problems that need to be addressed, and whether the occupants' needs have changed. It can also explore how effective the project delivery process was from concept to completion (Blyth *et al.*, 2006). Feedforward provides data for the design team to identify what can be learned for future projects, thus improving future building procurement. Finally, benchmarking provides a measure against which progress towards energy targets and sustainable construction can be compared (Menezes *et al.*, 2012). A POE thus assesses the technical performance (e.g. how well does the building fabric attain its design specification?), the functional performance (e.g. does the building sustain the occupants' and business' needs?) and the process (e.g. how well did the design and construction team perform?) (Blyth *et al.*, 2006).

During the mid-1990s, the first systematic POE of non-residential buildings was undertaken in the UK by the Post-occupancy Review of Buildings and their Engineering (PROBE) programme in an attempt to improve the openness of feedback to building services engineers (Cohen *et al.*, 2001). An initial study of 16 buildings, considered at that time (1995–1999) to be exemplar designs, found there appeared to be little connection between the predicted energy demand in the design estimations and computer models, and the actual measured demand post-handover. One of the main reasons identified was that the energy demand was often poorly specified in the briefing and design criteria (Bordass *et al.*, 2001).

Although POE is becoming more common for public buildings (although still unusual), there is less evidence that the same is occurring for homes and other non-residential buildings. This may be partly because it is more difficult to access occupants in their homes, particularly as an evaluation requires physical monitoring and/or the occupants to be interviewed or complete questionnaires. Obtaining a large enough sample to benchmark a development can also be a challenge (Stevenson and Leaman, 2010). It is also important for buildings that involve unconventional and unfamiliar technologies, which may lead to the occupants having to go through a period of adjustment, as we later highlight is the case with Passivhaus buildings (Section 4.10.3).

4.9.6 *The Soft Landings initiative*

A need to improve interactions and communications within and between the teams involved in the design and construction process, addressed by Construction Team Integration, is also a feature of the Soft Landings approach. In addition to this, the Soft Landings initiative is an attempt to improve the measured energy performance of buildings and reduce the energy performance gap, and to promote the continued involvement of the building contractors after the handover period. This can be achieved by making it a contractual requirement that they and the design team retain responsibility for the operation of the building for the first few years of occupation (Carbon Trust, 2012). The Soft Landings initiative also aims to encourage feedback between the teams involved. The scope is broader than just the relationship between contractor and occupant, but requires improved communication throughout the construction supply chain. In addition to the contractor retaining responsibility, it is hoped this 'learning by doing' approach will lead to deeper and more meaningful learning than that acquired by simply reading information (cf. Royston, 2014). The initiative identifies five key interlinked stages (BSRIA, 2014):

1   Inception and briefings. This details the expectations and responsibilities of the individual players, strengthening the relationship between the designer and the client, and defines the desired overall building performance.
2   Design development and review. The project team agree to adopt the Soft Landings approach, check the practicality of design solutions, review actual building performance and develop material for technical and user guides in consultation with the occupants.
3   Pre-handover. A team approach is required to ensure the building is operationally ready and the occupants' building management team understand the building's technical systems.
4   Initial aftercare. An aftercare team are available immediately after handover, usually on-site, to sort out any problems and answer questions from the occupants and their building management team.
5   Extended aftercare. The aftercare team make regular visits to review and fine-tune the performance of the building and provide guidance for its management. A number of independent post-occupancy surveys will also be undertaken (BSRIA, 2014).

For this approach to be successful, it is considered essential that lessons are learned regarding design, procurement and operation from previous projects. Occupants and building managers should be involved in the early briefing and design review stages to ensure effective communication and to provide opportunities to incorporate their views and requirements and to manage their expectations concerning building performance. It may also be necessary for Soft Landings requirements to be included in contract documents prior to the tender process (UK Green Building Council, 2013).

A Soft Landings approach was used to improve the handover process during the development and construction of Hackney City Academy. A series of pre-handover meetings were held to discuss how the school would be used, who would be trained and when, and the process to be used for taking over the buildings, including the process of bringing in the equipment and furniture. It was found that the move-in discussions were particularly important. Also, although the involvement of the facilities managers was useful, it was found that this needed to have been done earlier, with more thorough discussion, including the assignment of specific roles and responsibilities (Buckley, 2010).

Finally, it should be noted that although there seem to be numerous benefits to the Soft Landings approach, and it is being used for some public sector projects, it is unusual outside the public sector and has not yet been widely adopted. Few contractors seem to be prepared for it or know how to price it.

4.10 Case studies

4.10.1 Case study 1: Elizabeth Fry Building, University of East Anglia, Norwich

Rationale

This case study demonstrates several types of good practice, including: good team work; the use of POE and the use of a building energy management system (BEMS) to improve the efficiency of an already highly efficient building.

Project background

- Location: University of East Anglia, Norwich, UK
- Building type: educational and offices
- Tenure: owner-occupier
- Year of construction: 1995
- Size: 3,250 m^2
- Construction approach: Swedish hollowcore system called Termodeck, which consists of a floor slab that provides both a structural component and a method of ducting ventilation through the building, plus high-efficiency heat exchange units. The building has a high thermal mass, is relatively airtight (one air change per hour), with high levels of insulation (200 mm), plus low-E, argon-filled triple glazing (Standeven *et al.*, 1998).

When the Elizabeth Fry Building was constructed, it was thought to be the most airtight building in the UK (6.2 m^3/(h·m^2) at 50 Pa, reducing to 5.3 m^3/(h·m^2) at 50 Pa in 2011), with a total energy demand of half that considered to be good practice (Tofield, 2012). The building consists of two floors of lecture and seminar rooms, plus the top two floors of individual offices. It is heated by three 24 kW domestic condensing boilers with 50 per cent reserve capacity.

Approach

Prior to construction, the thermal performance of Elizabeth Fry was modelled to ensure that internal temperatures could be kept at acceptable levels even on hot days. In the summer the thermal mass of the building is cooled at night by air being passed through the hollowcore concrete slab (BRESCU, 1998). As part of the evaluation of the Elizabeth Fry Building, a POE was undertaken using a questionnaire survey.

Findings

The high level of thermal efficiency of Elizabeth Fry is attributed to good teamwork throughout the project. The client, the University of East Anglia, had a clear vision of its requirements and was prepared to put time and effort into ensuring this was communicated to the design and development teams (Standeven *et al.*, 1998). This good teamwork was extended to ensure the high standards of the design were translated into good practice during construction. What were then regarded as unusual features, such as triple glazing and minimal cold bridging, were carefully explained to site workers and were checked daily by the Clerk of Works, who worked closely with the contractor (Tofield, 2012).

Initially the building was not as energy efficient as predicted, with gas for heating usage of 65 kWh/m^2, due to an inadequate control system for monitoring and operating the systems. Although this energy demand was good for the UK at the time, being 20 per cent below the good-practice figure for academic buildings (Standeven *et al.*, 1998), it was not as efficient as Termodeck buildings in Sweden (Tofield, 2012). The original specification included a BEMS, but a 'value engineering' decision removed this to reduce costs (BRESCU, 1998). However, a year after handover the university installed a campus-wide BEMS, which enabled the building to be more easily and accurately controlled. This resulted in the gas use for heating being reduced by 50 per cent to 31 kWh/m^2 (Tofield, 2012).

The POE surveys indicated the building was popular, with its occupants scoring highly in terms of thermal comfort. Indeed, it gained the highest score in the Building Use Studies Comfort Index during the PROBE surveys (Standeven *et al.*, 1998).

Key messages

- The client was able to clearly communicate its vision to the design and development team.
- Good teamwork between the client, design and construction team contributed to a high-quality, energy-efficient building.
- An innovative low-energy ventilation system was used combined with exemplary airtightness.
- The installation of a campus-wide BEMS enabled further improvements to the energy management of the building.

4.10.2 Case study 2: Malmesbury Gardens social housing scheme, UK (Gupta and Dantsiou, 2013)

Rationale

This case study provides a detailed evaluation of the energy performance of a low-energy housing development, including the influence of construction practices at the design and construction stages, together with a number of technical aspects. The importance of effective communication is also demonstrated.

Project background

- Location: Swindon, UK
- Building type: residential
- Tenure: social housing
- Year of construction: 2011
- Size: 13 three- and five-bedroom homes; end- and mid-terrace
- Construction approach: hempcrete cast into a timber frame with exhaust air heat pumps to provide heating, hot water and mechanical ventilation. Each home was installed with 4 kWp PV and 4 m^2 solar thermal panels.

Approach

In an evaluation of Malmesbury Gardens, a Code for Sustainable Homes (CfSH) level five social housing development, Gupta and Dantsiou (2013) explored the influence of construction practices at the design and construction stages on the energy performance gap. In addition to a design and construction audit, the evaluation included fabric testing and a review of the commissioning process. The results were correlated with the occupants' experience post-handover.

Findings

There were key differences between the design intentions and the final performance. The homes were designed to a design standard of airtightness of 2 m^3/(h·m^2). A pressure test, however, gave an average airtightness of 3.3 m^3/(h·m^2). Overall, there was a 20 per cent difference between the design and 'as built' heat loss coefficient values. This was due to air leakage pathways, identified by thermographs, through badly fitted openings, skirting boards and the roof apex, in addition to inadequately sealed service pipes (Gupta and Dantsiou, 2013). The U-values of the hempcrete walls also did not meet expectations, with a final measured U-value of 0.47, compared to an anticipated design value of 0.18.

It was found that several changes in the design of the buildings had resulted in the energy performance (SAP) calculations not being updated. For example, the SAP calculations were for PVC-framed windows, but the specification notes were for timber-framed windows.

A review of the design specifications revealed several issues, such as the difficulty of accessing roof window handles because the PV panels had been fitted in the space immediately underneath the window. There were also issues with the usability of the air source heat pumps and the controls. Clarity was also a problem with the thermostat controls, as two different designs were used for the master and room thermostat; the interface also lacked clear labelling or any indication that the system had responded.

The main findings of the study were that the energy performance gap was due to a lack of effective communication between the design team, contractors and suppliers, together with the lack of proper system commissioning. There were also shortfalls in the handover process, and the challenges the contractors had when working with new technologies and materials. This latter finding related to issues concerned with the construction of the timber frame and the extended drying time of the hempcrete walls. The lack of effective communication was also considered to have led to many of the construction faults and incorrect system commissioning, while the lack of understanding of the systems resulted in the failing of the handover process to provide clear guidance to the occupants (Gupta and Dantsiou, 2013).

Key messages

- Newly constructed low-energy homes may still have an energy performance gap.
- Innovative energy-efficient construction methods and materials may not meet expectations.
- A lack of both effective commissioning and clear communication between the teams can lead to poor construction practices.
- An inadequate handover process, such as insufficient information about how to control their energy system, can leave tenants confused.

4.10.3 Case study 3: Wimbish Passivhaus scheme, Essex, UK (Ingham, 2013; 2014a)

Rationale

This case study provides illustrative examples of the use of the Passivhaus Planning Package (PHPP) and Standard Assessment Procedure (SAP) building modelling software. It also demonstrates the significance of occupation upon the energy performance gap, and provides a rare example of how models can sometimes over-predict building energy demand.

Project background

- Location: Tye Green, Wimbish, Essex, UK
- Building type: residential
- Tenure: social renters and shared-owners
- Year of construction: 2011
- Size: three blocks of 14 one-, two- and three-bedroomed homes, including flats (gross internal floor area of 950 m^2)
- Construction approach: masonry framework with external insulation (Passivhaus-certified).

This affordable housing development formed part of a UK Technology Strategy Board funded Building Performance Evaluation study, in association with Hastoe Group (the housing association leading the project). The development was the first rural affordable housing development in the UK constructed to the Passivhaus standard (see Chapter 6 for more details on the Passivhaus standard). It is highly regarded within Passivhaus circles, and won the best residential project in the 2012 Passivhaus Trust awards (Passivhaus Trust, 2012).

Approach

The energy performance data for this Passivhaus-certified housing development were collected through on-site energy monitoring equipment. These performance data were complemented by qualitative data (specifically, through interviews and participant observation), which focused on how the project was experienced. Both these sets of quantitative and qualitative data were sourced from discussions with and reports written by Ingham (2012, 2013, 2014a, 2014b) and colleagues.

Findings

Total actual electricity demand was found to be 61 per cent and 221 per cent higher than that estimated by PHPP and SAP, respectively, whereas heating fuel (gas) demand was found to be 20 per cent and 11 per cent lower than PHPP and SAP estimates (Figure 4.1).

A key reason for the energy performance gap being relatively small in this case study, in addition to actual heating fuel consumption being lower than both of the modelling estimates, is because of the efforts of contractors and building monitoring staff who invested considerable time and resources in helping the households adapt to Passivhaus technologies. In this regard, parallels can be drawn with the previous Elizabeth Fry case study (Section 4.10.1) and the Soft Landings approach (Section 4.9.6). Such resources were invested into the project because the key staff members had the desire and willingness to make the development a success. Moreover, since it was the first Passivhaus project for each member of the core project team, they were perhaps especially keen to interact with the households (including investigating teething problems and answering occupant queries), so as to learn as much as possible.

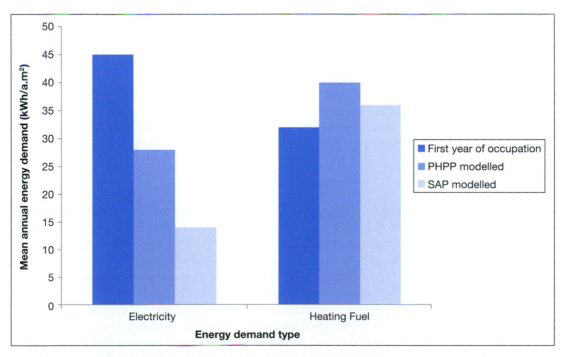

Figure 4.1 Comparison of the mean residential energy demand estimated by the Passivhaus Planning Package (PHPP) and Standard Assessment Procedure (SAP) software, with the mean actual energy demand during the first year of occupation (kWh/a·m²).

It is interesting to note how the SAP and PHPP clearly differ in how much energy they expected the Passivhaus homes to use. Key differences resulted from SAP focusing more on heating system controls, while PHPP delved more deeply into MVHR systems and the potential for solar gain (for more on comparisons between SAP and PHPP, see Reason and Clarke (2008)). We also note that SAP is not intended to be a design tool (e.g. to determine the size of central heating systems) and an accurate predictor of energy, despite it often being used for this purpose. Instead, SAP was designed to confirm compliance (or not) with UK building regulations.

While neither PHPP nor SAP is particularly better than the other at producing a more 'accurate' estimate of energy demand (although many will debate this), it is clear that both approaches have their own (often somewhat unusual) assumptions underpinning the specific representations of reality that each produced. For example, past versions of the PHPP software estimated the energy demand of electrical appliances by using assumptions for: the floor area; number of occupants (including proportions of occupants); and electricity demand of appliances. This estimated electricity demand is also used to calculate the internal temperature to determine whether or not the Passivhaus building will overheat. In addition, the PHPP modelling assumptions of appliances in this case were especially inappropriate, as they assumed the use of energy-efficient appliances (as per the Passivhaus standard requirements – see Section 6.2.3), which the social housing providers ended up not providing due to their associated maintenance obligations.

Key messages

- As with other case studies, it was found that energy performance gaps can occur even with highly energy-efficient buildings.
- Different models are based on different modelling assumptions, and thereby quantify the energy performance gaps differently.
- Design estimates will never be able to predict actual energy demand exactly, no matter how well optimised they may become.
- Post-occupancy expert involvement can help reduce occupational energy demand and, as such, minimise the performance gap, but it is not clear that this will always be possible.

4.11 Summary

We have seen in this chapter that there is often a significant difference between the predicted energy performance, as per the design of a building, and how that building actually performs in reality. This difference is termed the energy performance gap. There are numerous factors associated with this gap, spreading across the whole lifecycle of a building from conception to occupation. These factors include not only physical aspects associated with the buildings themselves, but also fragmentation of the design and construction process and how this leads to both problems of communication and a lack of integration between individuals and organisations. Significant variation in energy demand also occurs at the occupation stage, due to the unpredictable nature of how occupants use building technologies and hence demand energy. The rebound effect, which influences how people live (and demand energy) after the installation of an energy-efficient technology, can also mean that the expected energy performance of a building is not achieved.

Many of the initiatives developed to reduce the energy performance gap involve increasing integrated methods of working and enabling information flows between the various actors, including the provision of IT solutions. However, there is a danger in relying too much on energy performance models, which need to be considered carefully in the development of BIM systems.

For energy performance gap problems to be addressed and lessons to be learned, the occupation stage of a building's lifecycle needs to be included in any potential evaluation, and the possibility of contractors retaining some responsibility for the building post-handover needs to be explored. Furthermore, while challenging, we also need to think more carefully about why and how people seek to demand energy and use building technologies in the way that they do, because these influences should guide us when we seek to lower occupational energy demand. In addition, there is an argument that by improving our understanding of the underpinnings of energy demand, we may also start to re-frame our expectations regarding how much we can realistically reduce the performance gap, and indeed whether it can ever actually be closed.

References

ARUP (2013) The performance gap: Causes and solutions. Green Construction Board, Buildings Working Group. www.greenconstructionboard.org/images/stories/pdfs/performance-gap/2013-03-04%20 Closing%20the%20Gap_Final%20Report_ISSUE.pdf.

ARUP (2012) The building performance gap: Closing it through better measurement. National Measurement Network event report. www.npl.co.uk/upload/pdf/the-building-performance-gap.pdf.

Baiden, B.K., Price, A.D.F. and Dainty A.R.J. (2006) The extent of team integration within construction projects. *International Journal of Project Management*, 24(1), 13–23.

BIM Task Force (2013) FAQs. www.bimtaskgroup.org/bim-faqs, accessed 5 June 2015.

Blyth, A., Gilby, A. and Barlex, M. (2006) Guide to post occupancy evaluation. Higher Education Funding Council for England (HEFCE), London.

Bordass, B., Cohen, R. and Field, J. (2004) Energy performance of non-domestic buildings: Closing the credibility gap. In *Proceedings of IEECB '04 Building Performance Congress*, Frankfurt, Germany.

Bordass, B., Cohen, R., Standeven, M. and Leaman, A. (2001) Assessing building performance in use 3: Energy performance of the PROBE buildings. *Building Research and Information*, 29(2), 14–128.

BRESCU (1998) The Elizabeth Fry Building, University of East Anglia: feedback for designers and clients. Energy Efficiency Best Practice Programme, New Practice Final Report 106. www.termodeck. com/Filer/pdf/BRE_EFRY_REPORT.pdf.

BSRIA (2014) The Soft Landings Framework for better briefing, design, handover and building performance in-use. Building Services Research and Information Association, Bracknell, and the Usable Building Trust.

Buckley, M., Bordass, B. and Bunn, R. (eds) (2010) Soft Landings for schools case studies. Feedback from use of the Soft Landings Framework in new schools. BSRIA BG 9/2010, Bracknell.

Cabinet Office (2012) Government construction strategy: One year on report and action plan update. www.gov.uk/government/uploads/system/uploads/attachment_data/file/61151/GCS-One-Year-On-Report-and-Action-Plan-Update-FINAL_0.pdf, accessed 27 May 2015.

Cantin, R., Kindinis, A. and Michel, P. (2012) New approaches for overcoming the complexity of future buildings impacted by new energy constraints. *Futures*, 44, 735–745.

Carbon Trust (2012) Closing the gap: Lessons learned on realising the potential of low carbon building design. www.carbontrust.com/media/81361/ctg047-closing-the-gap-low-carbon-building-design.pdf, accessed 27 May 2015.

Cohen, R., Standeven, M., Bordass, B. and Leaman, A. (2001) Assessing building performance in use 1: The PROBE process. *Building Research and Information*, 29(2), 85–102.

DECC (2011) Low Carbon Building Programme (2006–2011) Final report, Department of Energy and Climate Change, London.

Demanuele, C., Tweddell, T. and Davies, M. (2010) Bridging the gap between predicted and actual energy performance in schools, *World Renewable Energy Congress XI*, Abu Dhabi.

Egan, J. (2002) *Accelerating Change, Report of Strategic Forum for Construction*, HMSO, London.

Egan, J. (1998) *Rethinking Construction: Report of the Construction Task Force*, HMSO, London.

Foulds, C. (2013) Practices and technological change: The unintended consequences of low energy dwelling design. PhD thesis, School of Environmental Sciences, University of East Anglia, Norwich, UK.

Fouquert, R. and Pearson, P.J.G. (2006) Seven centuries of energy services: The price and use of light in the United Kingdom (1300–2000). *The Energy Journal*, 27(1).

Gram-Hanssen, K. (2004) Domestic electricity consumption: Consumers and appliances. In: Reisch, L.A. and Røpke, I. (eds), *The Ecological Economics of Consumption*, Edward Elgar, Cheltenham.

Gram-Hanssen, K., Christensen, T.H. and Petersen, P.E. (2012) Air-to-air heat pumps in real-life use: Are potential savings achieved or are they transformed into increased comfort? *Energy and Buildings*, 53, 64–73.

Gupta, R. and Dantsiou, D. (2013) Understanding the gap between 'as designed' and 'as built' performance of a new low carbon housing development in UK. In *Sustainability in Energy and Buildings*, Springer, New York.

Guy, S. (2006) Designing urban knowledge: Competing perspectives on energy and buildings. *Environment and Planning C: Government and Policy*, 24(5), 645–659.

Guy, S. and Shove, E. (2000) *A Sociology of Energy, Buildings, and the Environment: Constructing Knowledge, Designing Practice*, Routledge, London.

HM Government (2012) *Building Information Modelling*, HMSO, London.

Hong, S.H., Oreszeczyn, T. and Ridley, I. (2006) The impact of energy efficient refurbishment on the space heating fuel consumption in English dwellings. *Energy and Buildings*, 38(10), 1171–1181.

Ingham, M. (2014a) Wimbish Passivhaus development: performance evaluation. Executive summary, Hastoe Group, London. www.hastoe.com/page/760/Wimbish-passivhaus-performs—Hastoe-releases-results-of-two-year-study-.aspx, accessed 5 June 2015.

Ingham, M. (2014b) Wimbish Passivhaus homes deliver. Hastoe Group, London.

Ingham, M. (2013) Wimbish Passivhaus: Building performance evaluation second interim report, Norwich. www.wimbishpassivhaus.com/WimbishInterimReport.pdf, accessed 5 June 2015.

Ingham, M. (2012) Wimbish Passivhaus: Building performance evaluation interim report. Build with CaRe, Norwich.

Innovation and Growth team (2010) *Low Carbon Construction IGT: Final Report*, HM Government, London.

Jevons, W.S. (1865) The coal question: Can Britain survive? In *The Coal Question: An Inquiry Concerning the Progress of the Nation, and the Probable Exhaustion of Our Coal-Mines*, A.W. Flux (ed.), Augustus M. Kelley, New York.

Madlener, R. and Hauertmann, M. (2011) Rebound effects in German residential heating: Do ownership and income matter? Institute for Future Energy Consumer Needs and Behaviour (FCN), RWTH Aachen. As referenced in Rosenow and Galvin (2013).

Menezes, A.C., Cripps, A., Bouchlaghem, D. and Buswell, R. (2012) Predicted vs. actual energy performance of non-domestic buildings: Using post-occupancy evaluation data to reduce the performance gap. *Applied Energy*, 97, 355–364.

Mills, E. (2011) Building commissioning: A golden opportunity for reducing energy costs and greenhouse gas emissions in the United States. *Energy Efficiency*, 4, 145–173.

Milne, G. and Boardman, B. (2000) Making cold homes warmer: The effect of energy efficiency improvements in low-income homes. A report to the Energy Action Grants Agency Charitable Trust. *Energy Policy*, 28(6), 411–424.

Monahan, J. (2013) Housing and carbon reduction: Can mainstream 'eco-housing' deliver on its low carbon promises? PhD thesis, University of East Anglia, Norwich, UK.

Monahan, J. and Powell, J.C. (2011) A comparison of the energy and carbon implications of new systems of energy provision in new build housing in the UK. *Energy Policy*, 39(1), 290–298.

Passivhaus Trust (2012) UK Passivhaus Awards 2012. www.passivhaustrust.org.uk/passivhaus_awards/uk_2012, accessed 27 November 2014.

Reason, L. and Clarke, A. (2008) Projecting energy use and CO2 emissions from low energy buildings: A comparison of the Passivhaus Planning Package (PHPP) and SAP. Association for Environment Conscious Building, Llandysul. www.aecb.net/PDFs/Combined_PHPP_SAP_FINAL.pdf, accessed 5 June 2015.

RIBA (2012) *BIM Overlay to the RIBA Outline Plan of Work*, Dale Sinclair (ed.). Royal Institute of British Architects, London.

Rosenow, J. and Galvin, R. (2013) Evaluating the evaluations: Evidence from energy efficiency programmes in Germany and the UK. *Energy and Buildings*, 62, 450–458.

Royston, S. (2014) Dragon breath and snow-melt: Know-how, experience and heat flows in the home. *Energy Research and Social Science*, 2, 148–158.

Shove, E. (2003) *Comfort, Cleanliness and Convenience: The Social Organization of Normality*, Berg, Oxford.

Sorrell, S. (2007) The rebound effect: An assessment of the evidence for economy-wide energy savings from improved energy efficiency. UK Energy Research Centre, London.

Sorrell, S. (2003) Making the link: Climate policy and the reform of the UK construction industry. *Energy Policy*, 31, 865–878.

Standeven, M., Cohen, R., Bordass, W. and Leaman, A. (1998) PROBE 14: Elizabeth Fry Building. *British Services Journal*, April, 37–41.

Steemers, K. and Yun, G.Y. (2009) Household energy consumption: A study of the role of occupants. *Building Research and Information*, 37(5–6), 625–637.

Stevenson, F. and Leaman, A. (2010) Evaluating housing performance in relation to human behaviour: New challenges. *Building Research and Information*, 38(5), 437–441.

Sunikka-Blank, M. and Galvin, R. (2012) Introducing the prebound effect: The gap between performance and actual energy consumption. *Building Research and Information*, 40(3), 260–273.

Tofield, B. (2012) Delivering a low-energy building: Making quality commonplace. Build with CaRe programme, University of East Anglia, Norwich.

TSB (2013) Retrofit revealed: The Retrofit for the Future projects. Data analysis report. Technology Strategy Board, Swindon.

UCL Energy Institute (2013) Summary of audits performed on Carbon Buzz. www.carbonbuzz.org/downloads/PerformanceGap.pdf, accessed 5 June 2015.

UK Green Building Council (2013) Pinpointing Soft Landings. Pinpointing discussion series. www.ukgbc.org/resources/additional/pinpointing-discussion-soft-landings, accessed 18 August 2015.

US Green Building Council (2014) LEED. www.usgbc.org/leed, accessed 27 May 2015.

Wang, L., Greenberg, S., Fiegel, J., Rubalcava, A., Earni, S., Pang, X., Yin, R., Woodworth, S. and Hernandez-Maldonado, J. (2013) Monitoring-based HVAC commissioning of an existing office building for energy efficiency. *Applied Energy*, 102, 1382–1390.

Winch, G.M. (2000) Institutional reform in British construction: Partnering and private finance. *Building Research and Information*, 28(1), 141–155.

Wingfield, J., Bell, M., Miles-Shenton, D., South, T. and Lowe, B. (2008) Lessons from Stamford Brook: Understanding the gap between designed and real performance. Report number 8, final report, Partners in Innovation Project: CI 39/3/663. www.leedsbeckett.ac.uk/as/cebe/projects/stamford/pdfs/stambrk-final-pre-pub.pdf, accessed 5 June 2015.

Zero Carbon Hub (2014) Closing the gap between design and as-built performance: Evidence review report. www.zerocarbonhub.org/sites/default/files/resources/reports/Closing_the_Gap_Between_Design_and_As-Built_Performance-Evidence_Review_Report_0.pdf, accessed 27 May 2015.

Zero Carbon Hub (2013) Closing the gap between design and as-built performance: Interim progress report. www.zerocarbonhub.org/sites/default/files/resources/reports/Closing_the_Gap_Bewteen_Design_and_As-Built_Performance_Interim_Report.pdf, accessed 27 May 2015.

5   Retrofitting buildings

5.1 Introduction

It is vitally important that we improve the energy efficiency of new buildings, but if countries are to meet their carbon targets the energy efficiency of existing buildings also needs to be substantially improved, in addition to reducing the carbon intensity of the energy supply. In the UK, for example, it is estimated that today's building stock will make up 80 per cent of buildings that will be standing in 2050 (Royal Academy of Engineering, 2010). These older buildings often have low levels of insulation and relatively high air leakage or natural ventilation rates, resulting in high energy consumption, carbon emissions and operating costs. In residential buildings substantial fuel bills can place a significant financial burden on households, especially those with a limited income. In addition to fuel bills for heating, many occupants also use a significant quantity of electricity for appliances, IT, lighting and ventilation. A proposed retrofit programme should include all uses of energy, not just heating, but in practice this rarely occurs apart from in commercial buildings. This chapter will examine the challenges posed by the urgent need to improve the energy efficiency of our current building stock.

Within the literature and across different countries, a variety of terms are used for the process of improving the energy efficiency of existing buildings, including retrofit, refurbishment, renovations and, in the USA, weatherisation. In this book we use the term 'retrofit'.

The type and extent of retrofitting needed to improve the thermal insulation of a building varies with its age, construction type and existing levels of insulation. For example, most buildings with cavity walls and lofts are relatively straightforward to insulate, although some cavities are 'hard-to-fill', meaning that the cavity is too small, difficult to access or shouldn't be filled due to structural (e.g. a metal frame construction) or locational reasons (e.g. significant exposure to wind-driven rain). Older buildings with solid walls (no cavity) and/or no loft (or loft access) present more of a technical challenge, and are more costly to retrofit.

It is often difficult to predict the amount of energy and carbon savings that will be realised from retrofitting. Modelling existing and retrofitted buildings is difficult as many details, such as U-values, are unknown. Also, the workmanship may be imperfect, and homes may be underheated prior to retrofitting. These issues are explored in more detail in Chapter 4, which discusses the energy performance gap.

This chapter explores several issues, including (1) how to fund retrofits; (2) whether it is better to demolish some older buildings and rebuild, rather than retrofit; (3) whether it is better for the most cost-effective individual measures to be undertaken in large numbers of properties or a 'whole house' approach be taken so that each property is improved to a high level of thermal efficiency; (4) how the decision to retrofit is made; and (5) if issues such as split incentives and disruption are barriers to retrofitting.

It is also important to consider that delivering energy-efficient buildings is not just an engineering issue. Physical systems are not independent of people's actions, be they specific building occupants or professionals in the industry. The interactions between physical buildings and the people who inhabit and retrofit them form a complex system (Institute for Sustainability, 2012) which is explored in more detail in Chapter 4.

Although the focus of this chapter is on residential buildings, we also explore the issues surrounding the retrofit of commercial buildings. In addition, we illustrate some of the retrofit challenges and issues using three case studies: Broadland Housing homes (Norwich, UK); homes in northern China; and public buildings in Serbia. Reports and databases from the UK Retrofit for the Future projects are also widely referred to because they include cutting-edge demonstration projects that were broadly monitored and evaluated (TSB, 2013a).

5.2 Age and efficiency of building stock

The energy efficiency of buildings is strongly related to their age due to the tightening of building regulations over time. For example, in Germany the average heating energy consumption is significantly lower for new buildings, with buildings constructed in 1918 currently using 250 kWh/m^2 for heating, whereas 2010 buildings use 53 kWh/m^2 (BPIE, 2011). Old, inefficient housing stock is a particular problem in Northern and Western Europe, with (in decreasing order) the UK, Denmark, Sweden, France, Czech Republic and Bulgaria having the oldest stock. Overall, approximately 40 per cent of European housing stock was constructed before the 1960s and urgently needs retrofitting (Economist Intelligence Unit, 2013).

In terms of carbon emissions per square metre, the current emissions of European buildings during occupation range from 5 to 120 kgCO$_2$/m^2, with an average of 54 kgCO$_2$/m^2 (BPIE, 2011). This can be compared with, for example, a target for new single-family residences in the UK of 17–20 kgCO$_2$/m^2 (BPIE, 2011). However, it is difficult to compare the carbon emissions of different countries due to differences in each country's calculation methods, fuel mix and the types of energy used in buildings. For instance, the carbon intensity of electricity in countries with high levels of nuclear, such as France, and hydroelectric power, such as Norway, is lower. The extent of adoption of district heating and co-generation also affects levels of emissions (BPIE, 2011).

If, instead of comparing carbon emissions, we compare the energy demand of European buildings, the effect of different fuel mixes is excluded (Figure 5.1). The energy demand will, however, still reflect the climate, and hence the heating/cooling demand, of individual

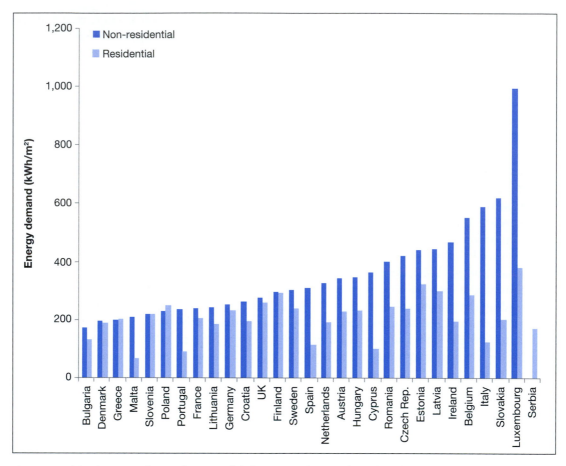

Figure 5.1 Total energy demand per useful floor area for residential and non-residential buildings in European countries (kWh/m²).

Source: Entranze database (accessed 2015).

Note: non-residential energy demand is not available for Serbia.

countries, in addition to the energy efficiency of the buildings. For example, in many hot countries, such as Italy, the high energy demand of non-residential buildings is likely to be associated with air conditioning. Figure 5.1 also illustrates the relative importance of residential and non-residential energy demand on a per square metre basis. Of course, the energy demand of buildings also relates to their size.

In the UK, 55 per cent of homes in 2011 were constructed before 1960, and 39 per cent before 1945 (National Refurbishment Centre, 2012). It has been estimated that only 1 per cent of the housing stock meets modern thermal efficiency levels (National Refurbishment Centre, 2012), with an average UK dwelling emitting nearly five tonnes of CO_2 per year (British Gas/Best Foot Forward, 2006). The UK's ageing building stock (Figure 5.2) is partly due to a low turnover rate of construction; only 8,000 dwellings are demolished and replaced each year (0.03 per cent of the current housing stock) (Ravetz, 2008).

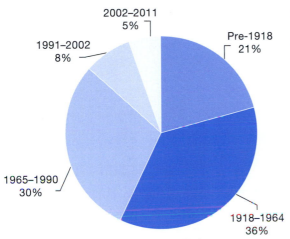

Figure 5.2
Proportions of UK housing stock by year of construction (%) (2011).

Source: Palmer and Cooper (2013, Table 4d).

Table 5.1 The number and proportion of hard-to-treat homes in England that are off-gas and/or have solid walls

	Both off-gas and solid wall	Off-gas only	Solid wall only	Total
Total	774,000	1,826,000	5,596,000	8,196,000
Proportion of hard-to-treat homes (%)	9	22	68	100
Proportion of all homes (%)	3.6	8.5	26.1	38.3

Source: Centre for Sustainable Energy (2011).

Some properties are relatively straightforward to retrofit, but older homes, in particular, can be more problematic. In the UK these homes, termed 'hard-to-treat', include solid-walled buildings, buildings with no loft space or no accessible loft space, high-rise flats and buildings that are not connected to the gas network (termed 'off-gas'). Off-gas properties are usually heated by electricity or oil at a significantly higher cost than gas, which is the main (91 per cent) heating fuel in the UK. Estimates indicate that there are approximately 8.2 million hard-to-treat homes in England, 38.3 per cent of the total housing stock (Centre for Sustainable Energy, 2011; Table 5.1). It has also been estimated that more than two-thirds of households living in hard-to-treat homes are in fuel poverty (Beaumont, 2007). This is based on the 10 per cent definition of fuel poverty, where more than 10 per cent of a household's income is spent on energy. (Fuel poverty is discussed further in Chapter 1.) Clearly there is the potential for retrofitting, to significantly reduce the extent of fuel poverty in these homes.

Over the last decade there has been a steady increase in the number of homes in the UK with cavity wall and loft insulation (Figure 5.3). It is estimated that 71 per cent of homes with cavity walls already have insulated cavities and 69 per cent of homes with lofts have at least 125 mm of loft insulation (Figure 5.4). Nevertheless, it has been estimated that to meet the UK government's 2050 carbon target, retrofitting will need to take place on an immense scale; 5,000 homes will need to be retrofitted per day until 2050 (National

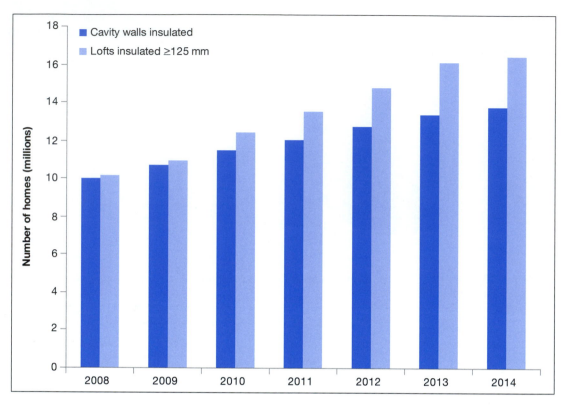

Figure 5.3 Homes in the UK with insulated cavity walls and lofts (≥125mm) (millions) (March 2008–March 2014).

Source: DECC (2014a).

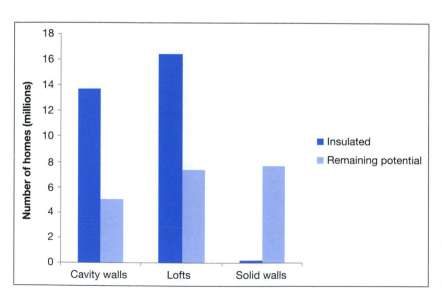

Figure 5.4 Homes in the UK with cavity wall, loft and solid wall insulation and those with remaining potential (millions) (2014).

Source: DECC (2014a).

Note: 'Remaining potential' includes homes that cannot physically be insulated or it is not cost-effective to do so. Homes where it is unknown if they are insulated are excluded.

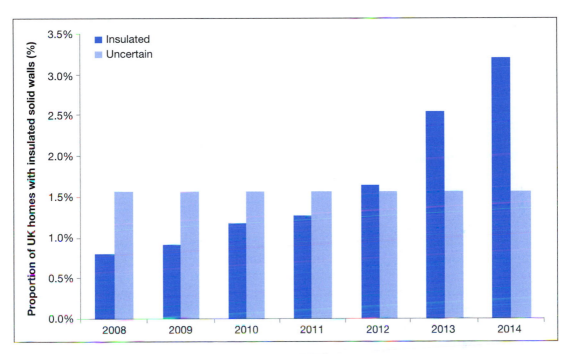

Figure 5.5 Proportion of UK homes with insulated solid walls (%) (March 2008–March 2014).

Source: DECC (2014a).

Note: 'Uncertain' indicates proportion of homes where it is uncertain if they are insulated or not.

Table 5.2 Typical U-values of different elements of three typical UK properties of different ages $(W/m^2 \cdot K)$

Element	Type/age of property	Element type/insulation level	Fabric U-value $(W/m^2 \cdot K)$
Roof	Period mid-terrace	Uninsulated	2.30
	1950s semi-detached house	Insulated 100 mm mineral wool	0.41
	2002 mid-terrace	Insulated 200 mm mineral wool	0.21
Walls	Period mid-terrace	Solid walls, 215 mm brickwork	2.10
	1950s semi-detached	Unfilled brick cavity	1.39
	2002 mid-terrace	Insulated brick/block cavity	0.35
Floor	Period mid-terrace	Uninsulated suspended timber floor	0.52
	1950s semi-detached house	Uninsulated suspended solid floor	0.57
	2002 mid-terrace	Insulated concrete	0.28
Windows	Period mid-terrace	Single-glazed timber frame	4.80
	1950s semi-detached house	Replacement U-PVC double glazing	3.10
	2002 mid-terrace	PVC double glazing	2.20
Typical leakage rate			**$(m^3/m^2/h$ at 50 Pa)**
	Period mid-terrace		12.00
	1950s semi-detached house		12.00
	2002 mid-terrace		10.00

Source: adapted from Energy Saving Trust (2010).

Refurbishment Centre, 2012). With only 3.2 per cent of homes with solid walls having solid-wall insulation (Figure 5.5), it is clear that the focus now needs to move towards retrofitting older properties if we are to address this massive challenge.

The relationship between age and efficiency is demonstrated in Table 5.2, which compares the fabric U-values of different elements for three UK properties of different ages and typical air leakage rates. A thermal element is a wall, floor or roof that separates the building from the outside environment or a part of the building that is heated to a different temperature (HM Government, 2010). A U-value is a measure of heat loss typically used for assessing the performance of thermal elements. The Energy Saving Trust (2010) identifies that unimproved period terrace properties offer significant opportunities for improvement.

5.2.1 Improving thermal insulation

Most European countries have thermal efficiency standards that need to be met when a building is retrofitted. To qualify for a loan in Germany, for instance, it is necessary for the primary energy demand of a retrofitted building to be no more than 115 per cent of the new-build legal maximum for space and water heating (Rosenow and Galvin, 2013).

In the UK, instead of an energy target for a building, the Building Regulations provide target U-values for the renovation of a thermal element in a building (Table 5.3). If a newly constructed thermal element is involved (e.g. an extension), more stringent U-values are required, especially for walls (0.28 W/m^2·K).

Older, unimproved houses, such as those with no loft insulation, are cost-effective to insulate, but as the depth of insulation is increased the rate of energy savings declines (the law of diminishing returns). For example, adding loft insulation to an unfilled loft will have a payback of less than two years, but the payback for adding additional insulation (from 100 to 270 mm) can be up to 14 years (Table 5.3). Some insulation measures, such as external wall insulation, can have payback periods over decades.

There are different methods of achieving these U-values. For example, external wall insulation can involve different types (mineral fibre or foamed plastic board insulation) and thicknesses of insulation (typically 50–250 mm), plus cement and sand or polymer rendering, or use cladding such as timber or metal (Construction Products Association, 2010). Although external wall insulation is more continuous than internal insulation, which has internal walls to contend with, there are detailing problems associated with external installation. For instance, where the wall joins the roof the additional wall width might result in the need for significant changes to the roof structure (Figure 5.6) (Galvin, 2011). Services, such as external pipes and guttering, may also need to be repositioned (Construction Products Association, 2010).

Insulation can be divided into relatively cheap but thick materials (e.g. mineral fibre) and thin, expensive materials (e.g. aerogel). To achieve high levels of insulation with the cheaper materials 400 mm is required in roofs and 250 mm in walls (Institute for Sustainability, 2012). To achieve high levels of insulation using internal wall insulation, without reducing the size of rooms, therefore requires more expensive materials to be used. The use of external wall insulation, particularly on the front of a building, can alter its appearance and may not be permitted if the building has heritage status. In some instances this can be partially overcome by using internal insulation on the front elevations and external insulation on the rear elevations that are less visible from the road.

Table 5.3 Summary of improvements to the thermal insulation of a dwelling required to achieve the UK target U-value (W/m²·K) (HM Government Building Regulations (2010) Part L1b) and the payback time (years)

Component	Starting point	Improvement Measure	Additional details	Approximate starting point U-value (W/m²·K)[a]	Target U-value (W/m²·K)[b]	Payback (years)[c]
Walls	Empty cavity	Fill with insulation	Mineral wool or polystyrene beads	1–1.60	0.55	<4
	Solid walls	Internal insulation	Dry-line walls with insulated plaster board[d]	2.30	0.30	12–18
		External insulation	Insulation board fixed to wall covered with cladding or rendered	2.30	0.30	19–27
Floors	Uninsulated solid/suspended	Insulation above floor	May require skirting boards raised and doors adjusted	1.00	0.25	~9
	Suspended	Insulation under floor	Insulation boards between joists/mineral wool suspended	1.30		
Ceiling/roof	Loft (empty)	0–270 mm 100–270 mm insulation	Mineral fibre quilt between rafters or blown insulation	2.30	0.16	<2 (2)[e]
	Loft with 100 mm			0.40		
	No loft space	Internal insulation	Insulation board between rafters or on sloping ceiling	2.30	0.18	<12 (>5)
Windows	Single	Replace with double/triple glazing	Gas-filled and/or low-emissivity coated glazing	4.80	1.60	
		Add secondary	Used in conservation areas where existing frames can't be replaced	2.7–3.10	2.40	
	Double	Upgrade				
	Frames	Insulated frames	Gas-filled and low-emissivity			
	DIY	Thick curtains Transparent film over window	Provides some improvement		1.60	<1
Doors		Insulated doors available	Improve existing doors using multi-point locks to reduce air leakage		1.80	
Draught proofing	DIY	Fill gaps around skirting boards and windows				~1

Source: adapted from Burton (2012); Construction Products Association (2010); Energy Saving Trust (2013).

(a) Construction Products Association (2010).
(b) HM Government Building Regulations (2010) Part L1b *Conservation of Fuel and Power in Existing Buildings*. The improved U-values are triggered when the area to be renovated is greater than 50 per cent of the surface area of an individual element, or 25 per cent of the total building envelope area.
(c) Energy Saving Trust (2013) (www.energysavingtrust.org.uk/Insulation, accessed 13 June 2013). Based on a three-bedroomed, gas-heated, semi-detached house.
(d) Fastened directly onto internal walls, or fastened to timber frame with insulation between timbers.
(e) Figures in brackets are for DIY provision.

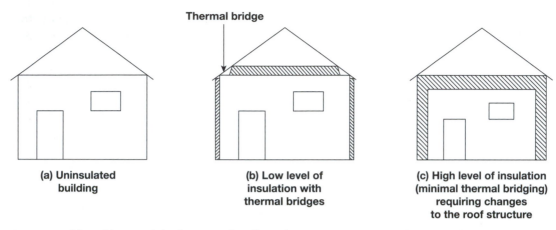

Figure 5.6 The addition of thick external wall insulation (c) to an uninsulated building (a) can result in the need for substantial changes to the roof pitch and eaves, or for the roof to be lifted if thermal bridges (b) are to be avoided.

Source: adapted from Galvin (2011).

5.2.2 Draught-proofing and ventilation

The thermal efficiency of a building can quickly be undermined by unwanted natural ventilation such as draughts through poorly fitting windows and doors that increase heat loss (see Chapter 2). Very low levels of air changes can nevertheless be achieved, and are required in buildings retrofitted to meet the Passivhaus standard (0.6 ach) (see Chapter 6). However, a 'tight' building with insufficient ventilation (and thereby insufficient circulation and replacement of internal air) can lead to poor air quality, and an unhealthy environment, with the accumulation of toxins and mould. There is some concern that retrofitting older homes, particularly, although not exclusively, those with solid walls, can also cause structural problems. In particular, increased airtightness with insufficient regard to ventilation can lead to condensation and damp, resulting in mould on thermal bridges and damp insulation, which will not perform well. To provide adequate ventilation in very airtight buildings, it is necessary to install some kind of ventilation system, mechanical ventilation with energy recovery often being the preferred choice. Ventilation is discussed in detail in Chapter 7.

Although retrofitting existing buildings is important, there are practical limits to how much energy can be saved through energy-efficiency improvements, due, for example, to thermal bridging. It has been estimated that the gas consumption of gas-heated homes can be reduced by up to half through insulating both the thermal envelope and hot water system (Institute for Sustainability, 2012). Further carbon savings require the installation of efficient heating systems, such as energy-efficient condensing gas-fired boilers or low-carbon technologies. The latter can include ground source heat pumps (GSHPs), air source heat pumps (ASHPs), biomass heating and solar thermal heating (usually for hot water).

5.2.3 Lighting and appliances

The largest use of energy in buildings is for space heating (around 70 per cent), but, as discussed in Chapter 1, significant quantities of gas and electricity are also used for hot water, cooking and appliances. The demand for electricity from European households and non-residential buildings has increased rapidly in the last two decades, at 1.7 per cent per year for households and 3.3 per cent per year for non-residential buildings (1990–2010) (Lapillonne *et al.*, 2012), reaching 29 per cent of their total energy demand by 2012 (EEA, 2013).

It has been estimated that electricity can be reduced by up to one-third through changing to energy-efficient lighting and appliances (Institute for Sustainability, 2012). For example, the phasing out of incandescent bulbs, which started in 2009, has led to a significant reduction in the energy demand of lighting, which will continue with the introduction of light emitting diodes (LEDs). Appliances and lighting are discussed further in Section 2.4.7.

5.3 Main stages of a retrofitting project

To undertake a retrofit project, five main stages are required (Ma *et al.*, 2012), although when applied to homes it is likely to be less formalised:

1   project setup and pre-retrofit survey;
2   energy audit;
3   the identification of retrofitting options;
4   site implementation and commissioning; and
5   validation and verification of energy savings.

In the project setup, the scope of the work is defined and targets are set if required. A pre-retrofit survey may also be undertaken if there are operational problems, so as to determine any technical issues and solutions, and identify any concerns of the occupants.

The purpose of the energy audit is to understand the energy performance of a building and its services (Ma *et al.*, 2012). Building energy data are collected and analysed, identifying inefficient energy use and developing potential energy conservation and efficiency measures. Energy audits vary in complexity, but three different levels have been identified (Pacific Northwest National Laboratory, 2011) that each build on the previous level. The level adopted will depend on the scope and goals of the retrofit, the size and complexity of the building, the degree of accuracy needed and the budget available.

1   A walk-through assessment that includes an assessment of energy bills and a brief site inspection. It identifies no- and low-cost energy-saving opportunities.
2   An energy survey and analysis that involves an in-depth analysis of energy costs, usage and building characteristics and identifies more expensive energy-saving opportunities.
3   A detailed energy analysis that, in addition to level 1 and 2 activities, involves monitoring, data collection and engineering analysis, and also provides a financial analysis for major capital investments.

The third stage of a retrofitting project identifies and compares potential energy-saving retrofitting options. This typically uses economic analysis tools, as well as energy- and risk-assessment models. During the fourth stage, the selected retrofitting measures are implemented and commissioned. For large buildings this will include a certain amount of 'fine-tuning' to ensure optimal operation.

The final, fifth, stage involves the verification and validation of the energy-saving measures. In addition to a technical assessment, this can include a post-occupancy survey to ensure the occupants and owners are satisfied with the results (Ma *et al.*, 2012).

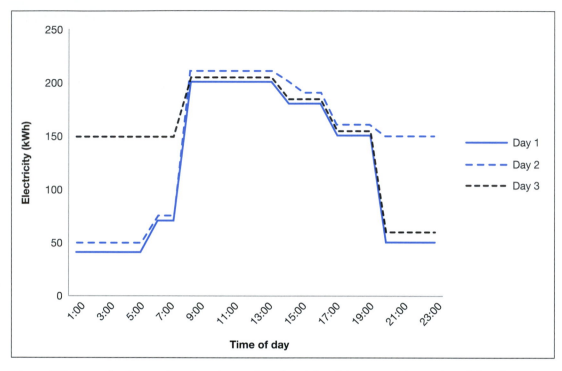

Figure 5.7 Example of precedent-based targeting; the night of day two and morning of day three show an unusually high level of electricity use (kWh).

Source: adapted from Carbon Trust (2012).

Ongoing monitoring is also important and can include precedent-based targeting, where current energy usage is compared with previous time periods (Figure 5.7), potentially prior to a retrofit. Activity-based targeting is also useful in that it relates energy consumption to its driving factors, such as manufacturing outputs or the weather, the variability of which can be normalised using degree days (Figure 5.8).

5.4 Retrofit challenges

Although the technologies and the stages required for retrofitting appear to be well established and the energy efficiency of buildings continues to improve in most Western countries, this is at a fairly low rate of increase. For example, the UK's housing stock improved by less than one Standard Assessment Procedure (SAP) point per year from 2001 to 2011 (Boardman, 2012). This gradual improvement has also been achieved, to some extent, by improving the thermal envelope of new buildings and undertaking the most cost-effective retrofit measures in existing buildings. If the efficiency of the building stock is going to continue to improve, less cost-effective and technically more demanding retrofit measures will need to be adopted. Several social, economic and technical retrofit challenges will need to be addressed if these measures are to be widely installed. This section explores these challenges by examining the following questions:

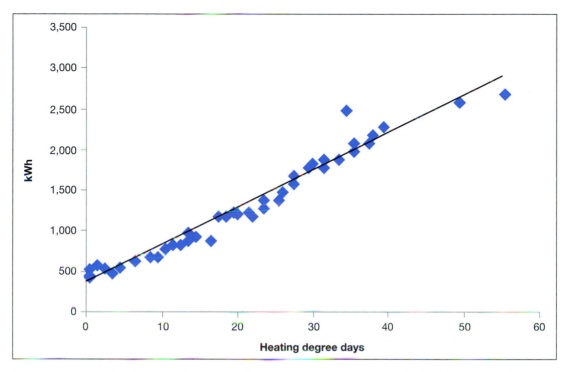

Figure 5.8 Example of activity based targeting where the gas demand of a building (kWh) has been plotted against heating degree days; when heating degree days are zero the gas could be used for water heating.

Source: adapted from Carbon Trust (2012).

- What are the upfront capital costs of retrofitting and what sources of funding are available to alleviate these costs?
- Is it more energy efficient to demolish old buildings and rebuild rather than to retrofit them?
- Is it more effective, both from a financial and a carbon perspective, to take a whole-building approach to retrofit or to improve more buildings using a measures-based approach?
- How do households make the decision to retrofit?
- How can the disruption to households during installation of a major retrofit be minimised? Is it better for occupants to move out?
- Do split incentives create a barrier to retrofitting?
- Is there a problem associated with a lack of skills and expertise, or a lack of trust in installers?

5.4.1 Cost and funding

The high cost of whole-house retrofitting can be a barrier to improving the thermal efficiency of existing homes. For example, Innovate UK's 'Retrofit for the Future' case studies estimated that it cost over £100,000 (€125,000) for each retrofit (Institute for Sustainability, 2012). An analysis of the costs is available in Sweett (2014). A whole-house approach involves the installation of a package of measures, including the building fabric, building services (heating, hot water, ventilation and lighting) and often renewable energy systems (Construction Products Association, 2014).

In addition, it is often necessary as part of retrofitting to include additional 'cosmetic' work, such as replacing kitchen units and windows, together with decoration, especially when internal solid-wall insulation is being undertaken. The cost of this additional work, plus the benefits from increasing thermal comfort by reducing draughts, and the health benefits from higher internal temperatures and reduced damp, are usually not accounted for.

Retrofitting can be much more costly when applied to individual or niche properties, rather than larger-scale projects involving numerous similar properties. The potential high cost of retrofitting is highlighted in the Broadland Housing case study (Section 5.7.1). The cost was particularly high due to the additional costs associated with improving one-off buildings rather than tackling this problem on a street-by-street basis as often occurs in other European countries. In the UK, with a high proportion of privately owned homes, this can be problematic. However, social housing estates, where many of the homes are owned by the same organisation and are of similar construction, are likely to be more cost-effective. The larger scale makes these retrofit projects more attractive to private finance, transaction costs are lower and the economies of scale and close proximity of similar properties makes the material and labour costs lower. Many countries also provide low-interest loans or grants to support social housing retrofits (Jankel, 2013).

In recognising the importance of retrofitting as a means to reducing both their national carbon emissions and fuel poverty, many governments provide financial support for energy efficiency measures. This support can be in the form of grants, loans, subsidies or a mixture of the three. The funds, which mainly come from two sources, central government and energy supply companies, are generally targeted at the fuel poor (Boardman, 2012).

5.4.2 New build verses retrofit

There has been considerable debate about whether it is more financially and environmentally effective to demolish old, inefficient buildings and rebuild new, or to retrofit old buildings (Ding, 2013). Retrofit is considered to be less damaging in terms of carbon emissions and resource use, as well as resulting in lower use of land and financial cost (Sustainable Development Commission, 2006). Retrofit is also thought to be more socially acceptable in that the community infrastructure is retained and existing communities are protected from disruption or even elimination (Power, 2008).

In the '40% House' project, Boardman *et al.* (2005) considered it will be necessary to demolish the three million most energy-inefficient homes (14 per cent of existing housing stock) and replace them with new energy-efficient homes if the UK is to meet its 2050 target of (at the time) 60 per cent reduction in carbon emissions. However, the UK Sustainable Development Commission (2006) did not consider this level of demolition likely, estimating that less than 10 per cent of the building stock will be demolished by 2050. Boardman *et al.* (2005) also did not take into consideration the embodied carbon required for the reconstruction, only the savings in operational carbon (Power, 2008); embodied carbon is the carbon emitted during the manufacture of products or generation of energy used during the process (see Chapter 3).

A study by the Empty Homes Services (2008), which compared the embodied carbon of constructing three new homes with the retrofitting of three homes, found that retrofitting resulted in an average of 15 tonnes of embodied CO_2 per home, while the construction of the new homes resulted in 50 tonnes of embodied CO_2 per home. Over an average lifetime the operational CO_2 emissions of the new-build homes was 10 per cent lower (174 tonnes CO_2) than the retrofitted homes (194 tonnes CO_2). When compared on a per metre squared basis, however, the emissions were similar for all six homes (1.7 tCO_2/m^2) because the retrofitted homes were larger. Over a 50-year timeframe, there was little difference between the emissions from the new-build and retrofitted houses. This would seem to indicate that retrofitted homes can be just as 'carbon efficient' as new build houses, but with a lower 'investment' of embodied carbon.

5.4.3 Whole-house vs. measures-based approach

Linked to the upfront cost of retrofitting is the question concerning whether it is better to improve the thermal efficiency of fewer homes to a very high standard (a 'whole house' or 'deep' approach) or to adopt a 'measures-based', 'elemental' or 'shallow' approach involving the installation of individual efficiency measures (or small-scale renewable energy technologies), often one at a time (Construction Products Association, 2010; Jones *et al.*, 2013). In the UK, many government-funded efficiency programmes, such as the Carbon Emissions Reduction Target, have used a measures-based approach which improves the thermal efficiency of more properties, but to a lower standard. One problem with this approach is that if only the most cost-effective improvements are undertaken initially, with a return visit at a later date to deliver further improvements, the cumulative cost is likely to be more expensive. A measures-based approach could consequently be less financially attractive in the long term, and the repeat visits more disruptive.

Although a whole-house approach is likely to be expensive and disruptive initially, the work is completed in one period of time. Moreover, significant reductions in heating costs and carbon emissions are achieved immediately (Construction Products Association, 2014). If it is possible to undertake a whole-house approach on a street-by-street basis, significant economies of scale can be achieved. For individual owner occupiers, a whole-house approach is often very expensive (Construction Products Association, 2014), with long payback times (see Section 5.4.1).

A whole-house approach was used by the Innovate UK's Retrofit for the Future programme, which aimed to achieve an 80 per cent reduction in CO_2 emissions in each of 100 homes. However, in an analysis of 37 of these homes, only three achieved an 80 per cent CO_2 reduction, with a further 23 achieving a 50–80 per cent reduction in CO_2 emissions (TSB, 2013a). The potential reasons for an energy performance gap are discussed in Chapter 4.

One problem with a whole-house approach is that the cost of retrofitting increases significantly with the percentage reduction in carbon emissions. An 80 per cent reduction in carbon emissions will cost in the region of £50,000 (€62,500) (Construction Products Association, 2014) or even more (see case study 1, Section 5.7.1). Multiple sequential measures to reduce emissions tend to follow the law of diminishing returns (i.e. the first efficiency measures installed save larger amounts of carbon than subsequent and future

Table 5.4 Levels of disruption from different types of retrofitting and the carbon cost-effectiveness of the measure; carbon cost-effectiveness ranges from 1 (>£500/tonne (€625/tonne)) to 5 (pays for itself within lifetime)

Disruption level	Example measures	Carbon cost effectiveness[a]
None	Low-energy lamps	5
	Low-energy appliances	2
Low (noisy or intrusive for a short time)	Heating controls	3
	Cavity wall insulation	5
	Loft insulation	5
Moderate (as above, but takes a little longer)	Replacement boiler	2
	Solar thermal heating	1
High (considerable disruption)	Replacement windows	2
	External wall insulation	4/5
Significant (will need to move out)	Internal wall insulation	5
	Ground floor insulation	5

Source: adapted from Construction Products Association (2014).

(a) The carbon cost effectiveness is the capital cost minus the fuel cost saving per tonne of CO_2 emissions saved during the measure lifetime.

measures), especially if the total energy savings are less than the sum of the individual measures (Jones *et al.*, 2013).

Individual measures vary in their carbon cost-effectiveness (i.e. the capital cost minus the fuel cost saving per tonne of CO_2 emissions saved during the measure lifetime) (Table 5.4). Some combinations of individual measures, such as improved insulation and a smaller but more efficient boiler, lead to more cost-effective energy savings (Jones *et al.*, 2013), while it is also more cost effective to reduce emissions by only 50–60 per cent (Construction Products Association, 2014). It also needs to be considered whether, once the most cost-effective measures are undertaken, a separate return visit is likely to be made to undertake the remaining, expensive, options. It must also be remembered that, if we are to meet our carbon targets, once the most straightforward homes to insulate are thermally efficient, it will still be necessary to tackle those that are hard-to-treat.

5.4.4 *Making the decision to retrofit your home*

In the UK, as in other Western countries, money spent on 'home improvements', such as new fitted kitchens, is seen as a positive way to improve the value of one's home (Wilson *et al*, 2013). Retrofit improvements, however, which are often hidden, are in general not perceived to add value to a home compared to a new bathroom or kitchen, and the potential financial savings and comfort benefits are often undervalued (Lewis and Smith, 2014). It is generally considered, therefore, that there is no link between the value of a building and its energy efficiency; the only financial benefit identified is the savings in energy bills (Boardman, 2012). A recent study, however, has identified a positive relationship between the energy rating and the price per square metre of a home, depending on its type and location (Fuerst *et al.*, 2013).

One way to overcome householder reluctance to undertake retrofitting is to link 'home improvements' and energy-efficiency measures. This can be done by encouraging homeowners to install energy-efficiency measures at the same time as other home improvements, such as installing a new bathroom or converting a loft space, or when moving into a new home (Construction Products Association, 2014). It is more cost-effective to do both together.

The decision to retrofit is generally thought to be influenced by the desire to save money, help the environment and improve comfort, but in general saving money has been found to have only a small influence on the decision-making process (Wilson *et al.*, 2013). In a study exploring how and why UK householders decide to renovate their homes, Wilson *et al.* (2013) found householders regard energy-efficiency measures in the same way as general amenity renovations – that is, as a means of changing aspects of domestic life that are creating tensions or issues within the home, such as changing available space within the home to make it more usable or comfortable. The decision to retrofit was also found to be influenced by what other people are saying and doing about renovations, and if the householder feels their home does not currently reflect their own sense of identity. Householders are considerably more likely to initiate energy-efficiency measures when the measures are part of broader renovations to their kitchens, bathrooms and living spaces (Wilson *et al.*, 2013). This study recommended that the Green Deal (see Section 5.5.2) could be targeted at householders planning to undertake general renovations, so the energy efficiency measures could essentially 'piggyback' onto the decision to undertake general renovations at minimal additional capital cost.

Research also highlights the important role of thermal comfort in people opting to retrofit (Maller and Horne, 2011). It appears that the decision to retrofit may have more to do with adequate warmth, and the social expectations attached to it, than actually saving energy. From this perspective, retrofitting has been considered merely as a means for households to adhere to the thermal comfort conventions that exist across a society (Shove, 2003).

5.4.5 Disruption and 'hassle'

One reason for not implementing energy-saving measures is thought to be the potential disruption to everyday life of major retrofitting projects, such as internal solid wall insulation and window replacement (Table 5.4). For example, internal wall insulation is likely to include plastering, plumbing and moving radiators and plug sockets. Occupants may also be apprehensive about the complexity of undertaking measures, in particular a whole-house approach, and concerned by the number of possible solutions. There may also be a fear of unknown structural problems being revealed during the retrofit process (National Refurbishment Centre, 2012) and the subsequent additional cost. In a UK study, 60 per cent of householders cited 'hassle' and their lack of knowledge as the main reasons for not undertaking energy-efficiency measures (Energy Saving Trust, 2011).

What we don't know is, if everything is perfect, with minimal or no disruption and the householders have all the knowledge they require, will they actually be any more likely to undertake retrofit measures? Many argue that filling a deficit in someone's knowledge does not guarantee action (e.g. Owens and Driffill, 2008; Hargreaves *et al.*, 2010), such as energy saving, or using low-energy technologies in the 'correct' way.

Many UK Retrofit for the Future (TSB, 2013a) projects found disruption during the retrofit (noise, dust, mess, having strangers in the home) was a problem for many of the residents who stayed in their homes during the retrofit. Moving residents out of their homes during the retrofit process is thus often regarded as the best way to minimise disruption to their everyday lives during the installation of technologies, although this can be just 'daytime decanting', when the residents vacate their homes during the day while the work is being undertaken. However, it can be difficult and expensive to find suitable accommodation for the duration of the retrofit, especially if the project overruns, which is common in many construction projects.

5.4.6 Split monetary incentives

A misaligned or split monetary incentive occurs in a transaction where the benefits do not accrue to the person who pays for the transaction. When applied to energy use in buildings, this occurs when the costs and benefits of improving the thermal efficiency of a building are not borne by the same person. Such might be the case, for example, when a building is owned by one person (the landlord) but occupied by another (the tenant). A split incentive can occur in a number of different situations:

*   Energy bills are included as a fixed amount in the rent, so the tenant has no incentive to reduce their energy consumption, such as by decreasing the internal temperature of the building or buying energy-efficient appliances. In this example, the tenant may experience the disruption of retrofitting but not benefit from lower bills.
*   The landlord of a building has no incentive to invest in energy-efficiency improvements when the tenant pays the energy bills because the tenant would reap the benefit of lower energy bills. This is a particular problem when a landlord is unable to increase rents after renovations due to legislative restrictions.
*   Developers retrofitting a building where market prices do not reflect the energy performance of the building.

The BPIE (2011) identified split incentives as a major barrier to the take-up of energy-efficiency measures. A comprehensive analysis by the IEA (2007) considered split incentives to be responsible for about 30 per cent of energy use. It is this barrier that underlies the rationale behind the UK's Green Deal (see Section 5.5.2).

5.4.7 Inadequate expertise

A lack of both management and construction expertise has been widely identified in a number of places. In the UK, 22 out of 54 TSB retrofit projects identified a lack of skills among the on-site workforce as a problem, with many recognising that the energy efficiency measures being installed were often new to the workers. Nine projects also found there were challenges concerned with management, in particular the need for close supervision of the on-site works (TSB, 2013a).

A UK survey of over 91,000 employers found 22 per cent of job vacancies were due to the shortages of skills, 23 per cent of which were in the construction industry (UK Commission for Employment and Skills, 2014). Of the construction establishments surveyed, 12 per cent had vacancies due to skills gaps and/or skills shortages. While construction establishments appear to be tackling this skills gap, with 56 per cent providing job training to 48 per cent of employees, this only amounted to an average of three days' worth of training per employee in 2013 (UK Commission for Employment and Skills, 2014). With the growing interest in low-carbon buildings, together with the use and installation of new materials and technologies, this does not appear sufficient.

During the economic downturn in the UK, many skilled workers had to leave the industry or retired. There was also a shift from permanent to short-term contracts due to the uncertainty surrounding the availability of continuous work. This resulted in less incentive for the construction industry to invest in training. With stronger economic growth in the industry, the TSB (2013a) suggested that developers improve and expand training for site operatives or that training should be provided by manufacturers. In addition, for complex technologies, manufacturers who also install their products should be selected to ensure a high-quality service and a responsibility for ongoing problems. It is hardly surprising that there is a significant skills shortage, because people often learn through doing (Royston *et al.*, 2014), and low-energy retrofitting has only started recently. The workforce can therefore be expected to gradually gain more skills as they are exposed to more and more retrofit projects. The problem is that it takes time, so the challenge is to shorten the learning process. New products, such as mechanical ventilation with heat recovery (MVHR), also need new knowledge, and poor practice is often undiscovered and unremedied (NHBC Foundation, 2013).

5.4.8 Other retrofit challenges experienced by project teams

The success of a retrofit project does not just depend on the expertise of its project team, but also on the wider institutional landscape of the sector. Problems with the supply chain are regularly experienced with, for example, 34 of the 54 Retrofit for the Future project teams experiencing a lack of choice or availability of products, often resulting in high prices and delays in delivery (TSB, 2013b). Product quality issues have also been identified with, for example, faults or not meeting the specified performance.

During construction projects of all types, unforeseen problems can occur that lead to additional work, delays and cost. This is particularly the case for retrofit projects where structural problems, such as the presence of bats, asbestos and wet rot, may not be identified until work on the building has started (TSB, 2013a). Retrofit projects can also experience space issues, including both the lack of space internally (e.g. for larger hot water cylinders) and externally (e.g. limits for external insulation due to need for bin storage) and space restrictions during the retrofit process itself (TSB, 2013a).

The following approaches and ideas have been developed to tackle some of these challenges in the future.

- To address product supply chain issues, project teams need to improve their knowledge of the supply chains so as to take into consideration availability and price volatility. This can be achieved by ensuring products are already in stock, by pre-ordering or by factoring in long lead times (TSB, 2013b). Although this approach goes some way to addressing the issues, a longer-term solution would be to help manufacturers and suppliers become more flexible and potentially to change their processes, rather than developers having to adapt to how the supply chain currently operates. For example, project developers can work collaboratively with suppliers to ensure warranties are available and valid (TSB, 2013b). (An example of collaborative working with suppliers is provided in Section 8.4.) Site visits can also be arranged for local suppliers to see how their products will be used, or for manufacturers to provide training in the installation and operation of their products.

- To inform and encourage buy-in from a project team and property occupants, a presentation of the aims, plans and timescale of a retrofit project can be organised before commencement of the work, ensuring that the timeframe is realistic and any disruption is not underplayed. This can be supported by additional training for on-site staff, potentially provided by product manufacturers or outside experts.

- Project teams can be frustrated when residents fail to adapt to their retrofitted homes or to adopt more efficient patterns of energy use (TSB, 2013b), although this may only be discovered if post-occupancy evaluation or monitoring takes place. There are many reasons why occupants fail to reduce their energy demand, some of which may not be obvious to the team. Project teams may need to reframe their expectations around how residents will use their new retrofit-related technologies, and acknowledge that there may well be difficulties (particularly initially) with the residents incorporating these innovative technologies into their lives. Residents may not use the technologies in a predictable way even if they are provided with information or if the project team seeks to 'educate' them. However, the provision of long-term, sympathetic support by, for example, well-informed housing association staff, may prove useful.

- Many suggestions were made by the Retrofit for the Future projects regarding cost overruns and delays. These included being realistic about the tasks and the risks, bearing in mind the lack of experience of some project teams and on-site staff. Forward planning and research into suppliers were identified as key, together with the need for a clear and detailed specification. Contingency planning is also important in order to understand the dependencies and complexities of the project (TSB, 2013b).

5.5 Policies to support retrofit

A range of different financing instruments, policy mechanisms and programmes are used globally to support retrofit, principally for fuel-poor households. These tend to vary depending on a country's economic circumstances, market development and extent of the fuel poverty problem (BPIE, 2011). In some countries, in addition to loans and grants, resources are provided for exemplar demonstration or kick-start initiatives. For example, in Europe the Zero Energy Building Renovation project, with a budget of €15.7 million, aims to demonstrate that a zero-carbon vision is technically, economically and socially feasible (Energy-efficient Buildings Public Private Partnership, 2013). The project includes raising the awareness of the public and public authorities, and identifying new business and management models. In Bulgaria, a joint initiative funded by the UNDP and the national government provides practical experience and half the costs for the renovation of multi-family residential buildings, with the owners paying for the remaining funding, about €2,000–3,000 for a medium-sized apartment (EC Eco-innovation Action Plan, 2011).

Here we provide two examples of policies supporting retrofit measures: the US Department of Energy (DoE) Weatherization Assistance Program (WAP), which aims to relieve fuel poverty and to stimulate the construction economy; and the UK Green Deal policy, which addresses the 'up-front' nature of the initial investment in energy-efficiency measures, together with the Energy Company Obligation (ECO) programme, which supports the fuel poor.

5.5.1 The US Department of Energy Weatherization Assistance Program

This ongoing programme, which started in 1976, provides grants to low-income households to improve the energy efficiency of their homes. Federal eligibility, which is defined as 60 per cent of state median income or below 150 per cent of the Federal Poverty Income Guidelines (whichever is higher), was estimated to apply to 35 per cent of US households in 2009 (Eisenberg, 2014).

The federal grants are paid to states and territories which organise contracts with local governments and non-profit organisations to provide retrofit services to those in need. Significant additional funding is provided by state and local agencies, Indian tribes, and the private sector. In the last evaluation of the State Energy Program, $540.9 million was shown to have been spent funding these programmes, of which $46 million came from the DoE. It has been estimated that this programme has helped reduce the energy bills of more than seven million households (Eisenberg, 2010). However, recent reductions in federal funding have resulted in fewer households being improved (State of Wisconsin, 2014).

In addition to insulation, this programme includes improvements to heating and cooling systems, electrical systems, household appliances, and health and safety aspects. As there is a cap on grants of $6,500 per household, the programme only funds cost-effective energy-efficiency measures. However, there have been calls to increase this grant cap, particularly for households with high energy needs.

An evaluation of the DoE programme estimated an average energy saving of 35 per cent in the first year, leading to an annual carbon saving per property of 2.65 tonnes. The average financial saving per property was $436 (€397) for heating and cooling, and $104–174 (€95–158) for electricity for lights and appliances, with a benefit:cost ratio for energy benefits alone of 1.8:1, assuming a 20-year lifetime of the measures (Eisenberg, 2010). A more recent evaluation of weatherisation programmes in Wisconsin state reported actual measured annual energy savings for single-family homes of 17 per cent for gas and 11 per cent for electricity, with greater reductions for mobile homes and multi-family dwellings (State of Wisconsin, 2014). The difference in the savings seems likely to reflect the inaccuracy of modelling energy saving.

5.5.2 The Green Deal and Energy Company Obligation

Two of the main barriers to improving the energy efficiency of buildings are considered to be the upfront capital costs of retrofit (see Section 5.4.1) and split incentives (see Section 5.4.6). In the UK, an attempt has been made to address these problems by the introduction, in 2013, of a financing mechanism called the Green Deal, which is available for both residential and non-residential buildings. The cost of the energy-efficiency improvements are paid for by the subsequent savings in the energy bills over time.

However, as we have seen earlier, the cost of retrofitting can be substantial, with some payback periods extending into decades. The Green Deal, therefore, established a 'golden rule' whereby the repayments of the loan must not exceed the expected financial savings. As a result, some retrofitting measures, such as external wall insulation, are unable to be funded using this mechanism. To take this, and the need to reduce fuel poverty, into consideration, an obligation (the Energy Company Obligation (ECO)) has been placed on energy companies to fund the necessary efficiency improvements (DECC, 2015a; Energy Saving Trust, 2014). The UK government is also currently looking at ways of applying the golden rule more flexibly or potentially scrapping it.

The measures included in the Green Deal comprise insulation, heating, draught-proofing, water heating, glazing and micro-generation. Lighting and mechanical ventilation with heat recovery measures are also included for non-residential buildings (DECC, 2015a).

The Green Deal process has four stages: (1) assessment, (2) finance, (3) installation and (4) payment. Using specifically designed software, a Green Deal assessor evaluates a building's energy use, providing an Energy Performance Certificate, in addition to determining what energy savings measures can be installed and estimating the typical energy savings likely to be achieved. They can also provide a quote for installing the measures, if required. A contract (the Green Deal Plan) provides details of the cost, repayments and interest rate. Once agreed, the plan can be put into action and the measures installed by a certified installer. To pay the initial cost, a loan is arranged with a consortium (the Green Deal Finance Company), with the repayments being taken directly from the energy bill.

The Green Deal is unusual in that the debt stays with the property rather than the building owner (Energy Saving Trust, 2014). If the home is sold, the new owner therefore has to take on the debt, which may be an obstacle to selling a property. The purchaser might consider that the thermal efficiency improvements to the home are reflected in the purchase price so they may not be prepared to also take on the loan. Thus a potential purchaser could require the repayment of the loan prior to purchase.

While the Green Deal provides loans for private homeowners, the ECO places an obligation on energy suppliers to financially support energy-efficiency measures for the fuel-poor and hard-to-treat properties. The ECO has three parts:

- The means-tested Affordable Warmth Obligation provides heating and insulation improvements for low-income and vulnerable households, although social housing tenants are not eligible.
- The Carbon Saving Target provides funding to insulate both solid-walled properties, with internal or external wall insulation, and properties with hard-to-fill cavity walls. It can be used in conjunction with the Green Deal to provide enough funding to make these relatively expensive measures cost-effective. However, in practice the Carbon Saving Target has not been successful at retrofitting large numbers of solid walls (Figure 5.9).
- The Carbon Saving Communities Obligation, which can be used to support the social housing sector, provides insulation measures to people living in the bottom 15 per cent of the UK's most deprived areas.

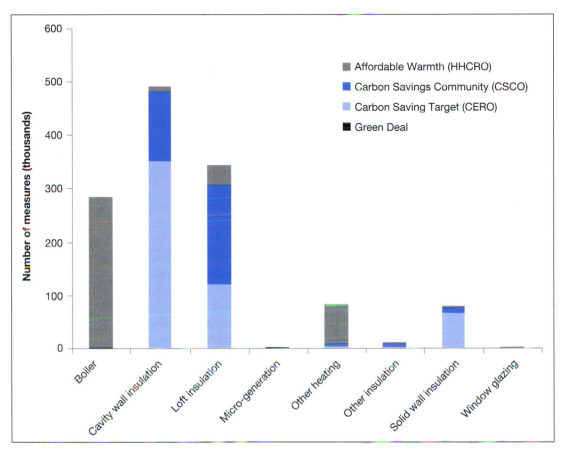

Figure 5.9 Number of ECO and Green Deal measures installed (thousands) (January 2013–December 2014).

Source: DECC (2015b).

Through the ECO, the UK government aims to help 230,000 low-income households or households in low-income areas to retrofit their homes. Energy suppliers are expected to invest £1.3bn (€1.63) per year, with a 75:25 split between the carbon and affordable warmth obligations.

Although the 'golden rule' should ensure that only cost-effective measures are funded through the Green Deal, the energy-saving assumptions are based on typical (thus averaged) rather than predicted or measured savings. As discussed in Chapter 4, it has been widely shown that estimated savings are often higher than actual measured savings (Rosenow and Galvin, 2013). It has been suggested that the Green Deal will only deliver about 26 per cent of the carbon reductions achieved by previous policies, the Carbon Emissions Reduction Target and the Community Energy Savings Obligation, mainly due to a reduction in the number of loft and cavity wall insulation measures which will not be compensated by an increase in solid-wall insulation (Rosenow and Eyre, 2012). These previous policies provided financial support to improve energy efficiency for all householders, compared with the Green Deal which is available to a smaller number of home owners. The Green

Deal may also be less attractive to households because of the commercial rates being charged on the loans, plus the unusual concept of attaching the debt to the home (Rosenow and Eyre, 2012).

By the end of December 2014, 1,337,000 measures had been installed in 1,086,000 properties through ECO, Cashback and Green Deal plans and the Green Deal Home Improvement Fund. However, 97 per cent of these were delivered through ECO. Although 447,309 Green Deal assessments were lodged up to the end of December 2014, only 8,348 households (1.87 per cent) had Green Deal Plans 'in progress' (including those installed) by the same date (DECC, 2015a). Of the Green Deal-funded measures, 26 per cent were for condensing boilers and 28 per cent for photovoltaic solar panels (Figure 5.10, which uses data for all boilers and micro-generation measures). The uptake of ECO has been far more popular, with 1,296,441 measures installed by the end of December 2014, 38 per cent for cavity wall insulation, 27 per cent for loft insulation and 22 per cent for boiler upgrades.

It would seem that the Green Deal is not very popular and is not living up to expectations. It has been criticised as being too complicated, the interest rates charged can be higher than commercial loans and it is not always certain that the money saved through energy savings will be sufficient to pay the loan repayments (Dowson *et al.*, 2012). Many home owners are also reluctant to take out a loan that remains with the property in case this puts off potential future purchasers. However, there is some evidence that people who have assessments go on to have retrofits but without Green Deal loans (DECC, 2014b).

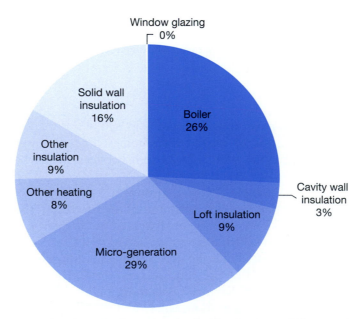

Figure 5.10 Proportion of each type of measure funded by Green Deal finance since commencement (%) (January 2013–December 2014).

Source: DECC (2015a).

5.6 Non-residential buildings

The energy demand of non-residential buildings represents a significant proportion of total energy demand in most developed countries. In the USA, commercial buildings are estimated to account for 18.4 per cent of total energy consumption (EIA, 2014). In the UK, commercial buildings are responsible for 10 per cent of the country's carbon emissions (Westminster Sustainable Business Forum and Carbon Connect, 2013), with nearly 70 per cent of these emissions arising from electricity use (Committee on Climate Change, 2013). The energy performance of these buildings can significantly deteriorate over time, with failings in the structure and systems. A US study identified over 100 types of faults that developed in commercial building systems, accounting for 2–11 per cent of the total energy consumption. Over one-third of these faults were due to things being left on while the building was unoccupied (Roth *et al.*, 2005).

Office buildings are often viewed rather differently than homes as they are an investment that needs profitable financial returns, whether they are owner-occupied or leased. The need for and extent of retrofitting is a trade-off between the retrofitting cost, improved building quality and environmental impacts (Juan *et al.*, 2009). However, any improvement in quality and environmental impact will usually need to be reflected in increased income to the owner, especially if the (lower) energy bills are paid by a tenant.

In the USA, as in many developed countries, the take-up of retrofit projects has been slow even though owners of commercial buildings can cost-effectively reduce their building's energy consumption. Building owners often consider energy retrofits to be too expensive despite empirical evidence indicating that the benefits are considerable and often profitable (Binkley, 2007). This is a particular problem for commercial buildings as retrofit returns are unique to each building and depend on contextual factors such as subsidies and electricity costs.

It has been estimated that UK businesses are missing out on cost-saving opportunities of up to £1.6 billion (€2 billion) through energy-efficiency investment (Westminster Sustainable Business Forum and Carbon Connect, 2013). This implies that decisions concerning whether to invest in energy-saving retrofits are complex, with many different aspects being taken into consideration. In particular, a lack of 'buy-in' from key individuals and organisations has been identified, both at a strategic and departmental level (Westminster Sustainable Business Forum and Carbon Connect, 2013). To further add to the complexity, the confidential nature of building contracts is likely to restrict access to commercial retrofit data (Benson *et al.*, 2011).

In terms of financial assessment, a survey of 100 firms found that the most commonly used decision-making rule for commercial buildings is the payback time (Harris *et al.*, 2000). Providing there is a good business case, the lack of funding is not considered to be a serious barrier, although organisations may lack the expertise to transform the identification of retrofit projects into viable business cases. It is, however, more likely to be organisational barriers within businesses that inhibit the provision of capital (Better Buildings Partnership, 2010). Split incentives can also be a problem for commercial buildings (see Section 5.4.6), particularly for those in multi-occupancy. Either senior management leadership or a clear procedure to invest in energy saving is required to give retrofit the necessary organisational priority (Better Buildings Partnership, 2010).

In energy terms, the effectiveness of a retrofit varies depending on a building's type, size and age, together with its geographic location, occupancy, energy sources, building fabric, operation and maintenance and services systems (Ma *et al.*, 2012). The most common retrofits for commercial buildings are controls, lighting and heating systems, because they provide a reasonable return on investments (ROIs) and are easily installed (Kok *et al.*, 2010).

5.7 Case studies

5.7.1 Case study 1: Broadland Housing (Broadland Housing, 2012)

Rationale

This case study illustrates a range of improvements used to improve the energy efficiency of homes using a whole-house approach and the energy and carbon emissions that can be saved. It also examines the embodied carbon of retrofit, and identifies the high cost of this holistic approach when applied to individual homes.

Project background

- Location: Norwich (UK)
- Building type: residential
- Tenure: social housing (Broadland Housing)
- Year of construction: 1800s to 2000
- Size: 11 homes of various sizes and ages
- Construction approach: retrofit
- Cost: £53,000–£85,000 (€66,250–€106,250) per home.

Broadland Housing (BH) provides 4,850 affordable social homes in East Anglia, UK. These include family homes, apartments, sheltered housing and housing with care (Broadland Housing, 2012). A pilot project was undertaken during 2011–2012 to refurbish 11 homes in Norwich (UK) to a high standard of thermal efficiency. It was funded by the European Regional Development Fund, and was designed to inform a future large-scale project to retrofit 1,000 BH homes.

A whole-house approach was taken that included high levels of thermal insulation, triple glazing, energy-efficient gas boilers and controls, as well as solar PV being installed on some properties (Table 5.5).

Approach

A detailed evaluation of the energy and carbon savings from retrofitting three of these homes was undertaken (Jones, 2012). Occupational energy use before and after the retrofit was calculated for each property using the Standard Assessment Procedure (SAP). In addition, actual energy use was determined through bill information, meter readings and some monitoring. From these data, carbon emissions and savings were calculated. The embodied carbon emissions were also calculated for the retrofit measures (Table 5.6).

Findings

The SAP calculations predicted a carbon saving from the efficiency improvements of 77 to 93 per cent (Table 5.6). Post-retrofit, the actual energy demand was lower than the modelled value, but the former was not calculated on a year-long basis at the time of the study.

Table 5.5 Summary of retrofit improvements made to homes studied in Broadland Housing Association retrofit project

	Home one		Home two		Home three	
	Pre-retrofit	*Post-retrofit*	*Pre-retrofit*	*Post-retrofit*	*Pre-retrofit*	*Post-retrofit*
Location	*City centre*		*City outskirts*		*Suburb*	
Construction year	*1800s*		*1980s*		*2000*	
Housing type	*Semi-detached*		*End terrace*		*Semi-detached*	
No. bedrooms	*3*		*4*		*2*	
Total floor area (m²)	*118*		*85*		*68*	
Walls	Solid brick, no insulation	50 mm insulation + insulating plaster	Cavity wall, poor-quality filling	50 mm insulation + insulating plaster	Cavity wall, poor-quality filling	External wall insulation, rendered
Roof	100 mm insulation	Completely sealed	150 mm insulation	300 mm insulation	100 mm insulation	Completely sealed
Floor	Suspended, no insulation	160 mm insulation across basement	Solid, no insulation	No change	Solid limited insulation	No change
Windows	Single glazed	Triple glazed, argon filled	Double glazed	Triple glazed, argon filled	Double glazed	Triple glazed, argon filled
Doors	No information	Insulated, timber frame at front, PVC at rear	No information	Insulated PVC doors, triple glazed	No information	Insulated composite, triple glazed PH
Heating	Gas boiler (65 per cent efficient)	Gas boiler with flue recovery (89 per cent efficient)	Gas boiler (78 per cent efficient)	Gas boiler with flue recovery (89 per cent efficient)	Gas boiler (76 per cent efficient)	Gas boiler with flue recovery (90 per cent efficient)
Heating controls	Programmer + room thermostat	Thermostatic radiator valve (TRV) with weather compensation	Programmer + room thermostat	TRV with weather compensation	Programmer + room thermostat	TRV with weather compensation
Hot water	Gas heating boiler	10 litre electric water heater	Gas heating boiler	Electric immersion + solar thermal	Gas heating boiler	Gas heating boiler + solar thermal
Lighting	100 per cent low energy	No change	33 per cent low energy	100 per cent low energy	20 per cent low energy	100 per cent low energy
Micro-generation	None	None	None	2.5 kWp PV + two solar thermal collectors	None	Two solar thermal collectors

Source: Jones (2012).

Table 5.6 Broadland Housing Association retrofit project predicted results: CO_2 and cost savings

	Home 1	Home 2	Home 3
CO_2 pre-works (kgCO_2/year)	8,735	3,434	2,651
CO_2 post-works (kgCO_2/year)	1,983	232	581
CO_2 saving (kgCO_2/year)	6,752	3,202	2,070
CO_2 saving (percentage of pre-retrofit emissions)	77.3	93.3	78.1
CO_2 post-works (kgCO_2/m²)	16.80	2.73	8.54
Embodied carbon (kgCO_2)	23,634	20,814	26,704
Payback time for CO_2 (years)	3.5	6.5	12.9
Cost of retrofit (£)	74,000	53,000	85,000
CO_2 saved in 20 years (tCO_2)	111	43	15
Cost of CO_2 saved over 20 years (£/tCO_2)	664.18	1,226.05	5,783.19

Source: Jones (2012).

The carbon embodied in insulation materials exceeded that embodied in both other materials and the transport of personnel and materials (Figure 5.11). All three homes had embodied carbon associated with their new heating boilers, solar thermal for homes two and three, and PV for home two. The all-round external wall insulation undertaken on home three resulted in the need for significant insulation materials, but this was partly

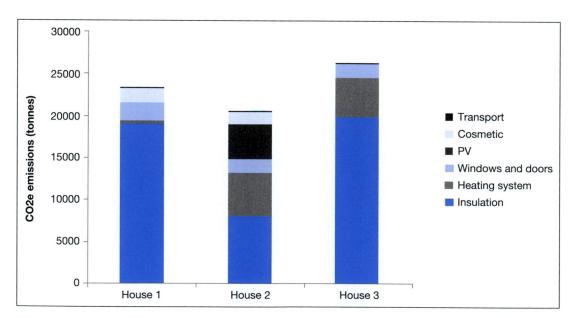

Figure 5.11 Embodied carbon emissions for retrofit BHS Housing (tonnes CO_2e).
Data source: Jones (2012).
Note: 'Heating system' includes boilers and solar thermal panels.

balanced by minimum internal alterations and hence little need for 'cosmetic' renovation. Although the materials used for cosmetic purposes did not add to the thermal efficiency, they were essential to reinstate the homes post-retrofit and are hence included in the calculations (Jones, 2012).

The financial cost of this retrofit was above average, mainly due to the high level of whole-house retrofit undertaken and the very different and individual approaches required for each house. The cost is much higher than the UK Green Deal budget of £10,000 (€12,500) per house. However, the additional cost of refurbishing hard-to-treat properties, such as home one, can be financially supported by the UK's ECO scheme (see Section 5.5.2).

Key messages

- Significant energy and carbon emissions can be saved using a whole-house approach.
- Retrofitting properties with low initial carbon emissions will result in low carbon savings and high payback times.
- A whole-house approach can be very expensive but may include substantial cosmetic improvements, which have additional value.
- Carbon is mainly embodied in the insulation materials. Although not used in this study, natural materials will have lower embodied carbon.

5.7.2 Case study 2: residential retrofit in China

Rationale

This case study illustrates the retrofit opportunities in China and the importance of the rebound effect in this emerging economy.

Project background

- Location: northern heating region, China
- Building type: residential
- Tenure: mixed
- Year of construction: various
- Size: various sizes and numbers of buildings
- Construction approach: retrofit
- Cost: four billion yuan (£0.4 billion; €0.5 billion) was allocated by Central Finance, plus an additional 12 billion yuan (£1.2 billion; €1.5 billion) from local government funds.

China is the world's highest energy user (22.4 per cent of global primary energy demand in 2013), overtaking the USA (19.8 per cent) in 2010 (BP, 2014). As in Western countries, the Chinese built environment is responsible for a substantial proportion of that demand and the subsequent carbon emissions. In 2007, the operational phase of the Chinese building sector was responsible for 31 per cent of the total Chinese energy demand and 17 per cent of total carbon emissions (Eom *et al.*, 2012). In 2005, China's per capita residential and commercial building energy consumption were 11.6 and 2.1 GJ, respectively

(China Energy Group, 2008), far less than Germany (31.1 GJ and 13.7 GJ) (Odyssee, 2009) and the USA (77.6 GJ and 64.1 GJ) (USBED, 2009). However, with its continuing economic growth and urbanisation, China's building-related energy demand is very likely to increase significantly in the future, thereby closing the difference that currently exists with Western countries (Eom *et al.*, 2012).

Of China's land area, 70 per cent consists of the northern 'heating region', where the heating demand for buildings predominates (Zhu, 2013). Within this area, 70 per cent of the buildings are considered to be energy inefficient (Shilei and Yong, 2009). Three main aspects of energy-inefficient buildings in the northern heating region have been identified: (1) building envelopes have poor thermal insulation performance, resulting in low indoor temperature in winter; (2) high air leakage rates, resulting in significant heat loss that accounts for nearly 30 per cent of the total supply heat; and (3) many homes have small, inefficient heating boilers (Yan *et al.*, 2011).

Approach

To address these challenges, the Heat Metering and Energy Efficiency Retrofit (HMEER) programme for northern heating regions of China was initiated in the 11th five-year plan (2006–2010), the aim being to retrofit 150 million square metres of existing residential buildings by 2010 (Yan *et al.*, 2011). To achieve this, a reward mechanism was adopted by Central Finance that took into consideration retrofit content, progress factor, finishing time and the climatic region: 55 yuan/m² (£5.50/m²; €6.88/m²) for severe cold regions and 45 yuan/m² (£4.50/m²; €5.63/m²) for cold regions (MOHURD, 2007).

Although the financial support for this large-scale project seems substantial, a complete retrofit typically costs 300 yuan/m², (£30/m²; €37.5/m²), so the total amount for the entire northern region would amount to 900 billion yuan (£90 billion; €112.5) (Shilei and Yong, 2009). Zhu (2013) found the initial investment for different combinations of retrofitting measures ranged from 2,287 yuan per building (£229; €286) (roof insulation) to 22,110 yuan (£2,211; €2,764) (roof and wall insulation, energy-efficient windows and a new heating boiler), with payback times ranging from 7 to 16 years.

Findings

Yan *et al.* (2011) considered the retrofit programme achieved five benefits:

1   The addition of thermal insulation and energy-efficient windows reduced heat loss, resulting in an increase of average indoor temperatures of 3–6 °C (from a pre-retrofit average temperature of 15 °C), thus improving thermal comfort without increasing the consumption of coal.
2   The thermal insulation of walls and windows reduced condensation fogging and mould, leading to a reduction in respiratory conditions and infectious diseases.
3   Energy-efficient windows had the additional benefit of improving sound insulation, providing the households with a quieter environment.
4   The lifetime of the improved buildings was prolonged and the value of each room was estimated to have increased by 500–1,000 yuan/m².
5   Retrofitting of the building facades improved the appearance of the buildings.

It was estimated that the HMEER programme saved 5.85 million tonnes of CO_2 in three years (Yan *et al.*, 2011). However, it is unclear if this takes into consideration the rebound effect mentioned above (the rebound effect is discussed further in Section 4.8.3). In a different study of rural residential buildings in the Chinese northern heating sector (Zhu, 2013), an average indoor temperature of only 8.6 °C was measured in January, which differs considerably from the 15 °C described above, although it is unclear what sample size this was based on. In addition, as part of the same study, a thermal comfort survey of 152 households indicated most occupants were, unsurprisingly, dissatisfied with the pre-retrofit indoor temperature of their homes (Zhu, 2013).

In order to explore the retrofitting opportunities for Chinese buildings in more detail, Shilei and Yong (2009) examined the retrofitting of three residential buildings constructed in 1978. Prior to retrofit, the maximum indoor temperatures were found to be 16 °C, with obvious heat bridges from external walls, and extensive condensation and mildew on internal walls. Most of the steel windows would not close and the fly ash roof insulation was damp. The thermal envelope was significantly improved by insulating the walls, doors, windows, roof, staircase and partition walls of the basement. Insulation was also applied to the heating supply system and heat metering equipment was installed. Post-retrofit, the internal temperature increased to 20 °C and the heating energy consumption had been reduced by 30 per cent (Shilei and Yong, 2009).

These findings indicate there is likely to be a significant level of rebound with thermal improvements, particularly in rural homes as residents are able to increase the temperature of their homes, thus improving their quality of life.

Following the success of the HMEER programme, China's twelfth five-year plan (2011–2015) aims to retrofit a further 400 million square metres of existing residential buildings in the northern region (Yan *et al.*, 2011).

Key messages

- There are significant opportunities to improve the thermal efficiency of Chinese homes.
- Many Chinese homes in the northern heating region are currently underheated, leading to high levels of rebound.
- Although this large-scale retrofit programme required substantial investment, with fairly long payback times, the improvements in thermal comfort, appearance and financial value are considerable.

5.7.3 Case study 3: retrofit of public buildings in Serbia (Bećirović and Vasić, 2013)

Rationale

This case study illustrates the application of a clear methodological approach to the evaluation of energy savings from a retrofit programme.

Project background

- Location: Serbia
- Building type and tenure: public buildings

- Year of construction: not stated, but many buildings constructed in the 1970s and 1980s
- Size: 62 buildings: 28 schools, 29 health care institutions and 5 social care institutions
- Construction approach: retrofit.
- Cost: €11.6 million in total, 47 per cent on schools, 47 per cent on health care institutions and 6 per cent on social care institutions; payback is expected in 13 years.

Inefficient energy use is of major concern in Serbia, where primary energy demand per dollar of GDP is considerably higher than many EU countries, which is (in 2012) 13 times higher than Germany and ten times higher than France (Bećirović and Vasić, 2013). The building sector is expected to contribute to energy-efficiency plans by reducing its energy demand by 9 per cent by 2018 (Todorović, 2010). There is considerable potential to meet this reduction (Šumarac, 2010) as many buildings were constructed of reinforced concrete and bricks with no thermal insulation. In addition, wood and metal windows and doors have deteriorated and heating systems are often in a poor condition.

The Serbia Energy Efficiency Project (SEEP), which was implemented in two phases, targets public facilities such as schools, health care and social care institutions. A summary of the methods and findings of the second phase (2009–2012), which are provided here, are presented in detail in Bećirović and Vasić (2013).

Approach

In line with procedures proposed by the Buildings Performance Institute Europe (BPIE, 2013), a step-by-step procedure was used to select the energy-efficiency improvements to be implemented. This included the following:

- An energy audit; each building was monitored pre- and post-retrofit.
- A model of the building's energy consumption.
- Identification of energy-efficiency improvements (Table 5.7).

Table 5.7 Energy-efficient measures applied in retrofitted buildings in the second phase of the Serbia Energy Efficiency Project

Measure	Number of buildings where measure applied
Facade joinery replacement	49
Facade wall insulation	23
Roof insulation	43
Installation of thermostatic radiator valves	43
Installation of balancing valves	13
Installation of variable flow pumps	4
Modernisation of boiler room	15
Fuel switch	1

Source: adapted from Bećirović and Vasić (2013).

- A computer simulation of energy savings for each improvement (using RETScreen 4).
- Identification of feasible energy-efficiency packages (2–5 packages for each building).
- A computer simulation of energy savings attributed to each proposed energy efficiency package (RETScreen 4).
- Identification of cost-optimal energy.
- A decision by stakeholders on which energy efficiency package to implement.

Findings

The average gross final energy demand was reduced by 47 per cent, with the average energy demand on an area basis changing from 302 kWh/m^2 to 159 kWh/m^2. The average energy saving for schools was 49 per cent, which is particularly impressive given that the pre-retrofit indoor temperatures of classrooms were 14–15°C. Unfortunately, the post-retrofit temperatures are not provided, but it is clear that a significant level of rebound occurred. The final energy demand of health care institutions was reduced by 47 per cent, while that of the social care institutions was 41 per cent and the Clinical Centre of Nis by 54 per cent. The energy savings could have been larger, but the majority of the buildings only had a few energy-efficiency measures installed, mainly due to budgetary constraints, and the project works were restricted as many buildings had to remain operational during the retrofit process. The initial condition of the buildings was also very poor and as a consequence some construction improvements were required that were not associated with energy savings.

Key messages

- Significant energy savings can result from investment in retrofit programmes for public buildings.
- Despite high energy costs, payback times are high partly due to the initial poor condition of the building, requiring addition construction work.
- If buildings have to remain operational during retrofit it may restrict the measures undertaken.

5.8 Summary

There is an important and urgent need to improve the energy efficiency of our current building stock. Energy efficiency is strongly influenced by a building's age, with older, inefficient housing stock being a particular problem in Northern Europe. Older buildings, especially those with solid walls, are more difficult and expensive to retrofit, but do have the potential for greater energy savings. While unwanted ventilation is a particular problem in old or poorly constructed buildings, which can quickly undermine the energy efficiency of a building, some ventilation is essential to maintaining healthy air quality.

There is considerable debate as to whether a 'whole-house' approach (aiming to reduce emissions by 80 per cent) or a measures-based approach should be taken. The Broadland Housing case study (Section 5.7.1) demonstrates the high cost of a whole-house approach, particularly for older homes. However, the one-off nature of the properties included in this project is likely to have made them more expensive than if several similar homes were retrofitted at the same time. Although some combinations of measures can lead to improved savings, in general the cost of retrofitting increases dramatically with the percentage reduction in carbon emissions; a more cost-effective 'middle ground' approach could be to reduce emissions by 50–60 per cent.

The question has also been posed as to whether it is better to retrofit old buildings or demolish them and rebuild. In terms of net carbon emissions, it generally seems better to retrofit, although there are likely to be instances where buildings are in too poor a condition. Although retrofit can cause some temporary disruption to the day-to-day lives of households, the social disruption associated with demolition and the construction of new homes can be far more significant.

While there have been many retrofit projects that have successfully reduced energy demand, the energy savings are generally less than anticipated. Modelled predictions can be inaccurate, workmanship may be less than perfect and thermal bridging hard to avoid. Homes may also be less 'leaky' than anticipated or, if the occupants underheat their home prior to the retrofit, the anticipated financial savings will be less due to the rebound effect. However, as demonstrated by the case studies, even though the Chinese homes and Serbian public buildings were underheated prior to retrofitting, substantial energy savings can be realised, especially in buildings that were originally of poor-quality construction. In addition to the energy savings, the Chinese case study (Section 5.7.2) also reveals other improvements to the internal environment, such as significantly reduced damp conditions and improved sound insulation.

The retrofitting of non-residential buildings is similar to that of homes, but it can lead to different challenges due to the way in which the institutions are organised, such as the expectations of commercial retrofit projects to generate a profit. It is important, for example, for landlords to be able to increase the income from the property in order to repay their investment.

The barriers to retrofitting are numerous and complex. Many consider the cost of retrofitting can inhibit the uptake of measures, particularly as the occupants of old or poorly constructed buildings are least likely to be able to afford the necessary improvements. Many governments, therefore, provide financial support, often as direct subsidies but also by introducing schemes such as the Green Deal in the UK, which aims to improve access to loans. The Green Deal is also designed to address the issue of split incentives, where the capital cost of thermal improvements are paid for from the energy savings. However, the take-up of Green Deal loans has been very low, revealing that the challenge to get people to retrofit their homes is far greater than overcoming financial barriers.

Indeed, as demonstrated by the Retrofit for the Future projects, there are current challenges associated with a lack of skills and expertise on retrofitting, especially when a high quality of construction is required, as in the case of a Passivhaus standard. This lack of experience not only concerns the construction process, but also applies to procurement and supply chains, micro-generation technologies and management practices.

References

Beaumont, A. (2007) Hard-to-treat homes in England, *W07 – Housing Regeneration and Maintenance International Conference*, 25–28 June, Sustainable Urban Area, Rotterdam, Building Research Establishment.

Bećirović, S.P. and Vasić, M. (2013) Methodology and results of Serbian energy-efficiency refurbishment project. *Energy and Buildings*, 62, 258–267.

Benson, A., Vargas, E., Bunts, J., Ong, J., Hammond, K., Reeves, L., Chaplin, M. and Duan, P. (2011) Retrofitting commercial real estate: Current trends and challenges in increasing building energy efficiency. UCLA Institute of the Environment and Sustainability. www.environment.ucla.edu/media/files/Retrofitting-Commercial-Real-Estate-30-mlg.pdf, accessed 28 May 2015.

Better Buildings Partnership (2010) Low carbon retrofit toolkit: A roadmap to success. www.betterbuildingspartnership.co.uk/sites/default/files/media/attachment/bbp-low-carbon-retrofit-toolkit.pdf, accessed 18 August 2015.

Binkley, A.G. (2007) Real estate opportunities in energy efficiency and carbon markets. Massachusetts Institute of Technology. http://dspace.mit.edu/bitstream/handle/1721.1/42034/228656560-MIT.pdf?sequence=2, accessed 28 May 2015.

Boardman, B. (2012) *Achieving Zero: Delivering Future-friendly Buildings*, Environmental Change Institute, Oxford.

Boardman, B., Darby, S., Killip, G., Hinnells, M., Jardine, C.N., Palmer, J. and Sinden, G. (2005) *40% House*, Environmental Change Institute, Oxford.

BP (2014) *BP Energy Outlook 2035*, British Petroleum, London.

BPIE (2013) A guide to developing strategies for building energy renovations. www.bpie.eu/documents/BPIE/Developing_Building_Renovation_Strategies.pdf, accessed 28 May 2015.

BPIE (2011) Europe's buildings under the microscope: A country-by-country review of the energy performance of buildings. www.bpie.eu/eu_buildings_under_microscope.html#.VWcF1M9VhBc, accessed 28 May 2015.

Best Foot Forward (2006) Green cities report: domestic carbon emissions for selected cities, report for British Gas.

Broadland Housing (2012) Annual review. http://broadlandgroup.org/uploads/publications/annual_review/Annual%20Review%202012.pdf, accessed 5 June 2015.

Burton, S. (2012) *Handbook of Sustainable Refurbishment*, Earthscan, Abingdon.

Carbon Trust (2012) Monitoring and targeting. www.carbontrust.com/resources/guides/energy-efficiency/monitoring-and-targeting, accessed 5 June 2015.

Centre for Sustainable Energy (2011) Analysis of hard-to-treat housing in England. Internal research paper, Bristol.

China Energy Group (2008) China Energy Databook v7.0. https://china.lbl.gov/research-projects/china-energy-databook, accessed 18 August 2015.

Committee on Climate Change (2013) Meeting carbon budgets – 2013 progress report to Parliament. www.theccc.org.uk/publication/2013-progress-report, accessed 28 May 2015.

Construction Products Association (2014) *An Introduction to Low Carbon Domestic Refurbishment*, 2nd edn, Newcastle-Upon-Tyne, RIBA.

Construction Products Association (2010) *An Introduction to Low Carbon Domestic Refurbishment*, 1st edn, Newcastle-Upon-Tyne, RIBA.

DECC (2015a) Domestic Green Deal and Energy Company Obligation in Great Britain, monthly report, February, Department of Energy and Climate Change, London.

DECC (2015b) Green Deal and ECO database, February 2015, Department of Energy and Climate Change, London.

DECC (2014a) Energy consumption in the UK (ECUK): Domestic data tables, Department of Energy and Climate Change, London.

DECC (2014b) Green Deal and Energy Company Obligation and insulation levels in Great Britain, quarterly report, September, Department of Energy and Climate Change, London.

Ding, G. (2013) Demolish or refurbish: environmental benefits of housing conservation. *Australian Journal of Construction Economics and Building*, 13(2), 18–34.

Dowson, M., Poole, A., Harrison, D. and Susman, G. (2012) Domestic UK retrofit challenge: Barriers, incentives and current performance leading into the Green Deal. *Energy Policy*, 50, 294–305.

EC Eco-innovation Action Plan (2011) Renovating multifamily buildings in Bulgaria. http://ec.europa.eu/environment/ecoap/about-eco-innovation/good-practices/bulgaria/706_en.htm, accessed 5 June 2015.

Economist Intelligence Unit (2013) Investing in energy efficiency in Europe's buildings. A view from the construction and real estate sectors. www.bpie.eu/uploads/lib/document/attachment/15/EIU_GBPN_EUROPE.pdf, accessed 28 May 2015.

EEA (2013) Total final energy consumption by sector in the EU-27, 1990–2010. www.eea.europa.eu/data-and-maps/figures/final-energy-consumption-by-sector-6#tab-european-data, accessed 5 June 2015.

EIA (2014) Monthly energy review. www.eia.gov/totalenergy/data/monthly/index.cfm?src=Total-b4#consumption, accessed 5 June 2015.

Eisenberg, J.F. (2014) Weatherization Assistance Program technical memorandum: Background data and statistics on low-income energy use and burdens. Oak Ridge National Laboratory, Oak Ridge.

Eisenberg, J.F. (2010) Weatherization Assistance Program technical memorandum: Background data and statistics. Oak Ridge National Laboratory, Oak Ridge.

Empty Homes Services (2008) New tricks with old bricks. www.emptyhomes.com/empty-homes-publications-and-toolkits/empty-homes-publications, accessed 28 May 2015.

Energy-Efficient Buildings Public Private Partnership (2013) EeB PPP Project Review: FP7-funded projects under the 2010, 2011 and 2012 calls. http://ec.europa.eu/research/industrial_technologies/pdf/eeb-ppp-project-review-2010-2011-2012_en.pdf, accessed 28 May 2015.

Energy Saving Trust (2014) Green Deal information. www.energysavingtrust.org.uk/scotland/domestic/improving-my-home/green-deal, accessed 5 June 2015.

Energy Saving Trust (2013) Home insulation. www.energysavingtrust.org.uk/domestic/content/home-insulation, accessed 5 June 2015.

Energy Saving Trust (2011) Trigger points: A convenient truth – promoting energy efficiency in the home. www.energysavingtrust.org.uk/Publications2/Corporate/Research-and-insights/Trigger-points-a-convenient-truth, accessed 28 May 2015.

Energy Saving Trust (2010) Sustainable refurbishment. http://tools.energysavingtrust.org.uk/Publications2/Housing-professionals/Refurbishment/Sustainable-Refurbishment-2010-edition, accessed 5 June 2015.

Eom, J., Clarke, L., Kim, S.H., Kyle, P. and Patel, P. (2012) China's building energy demand: Long term implications from a detailed assessment. *Energy*, 46(1), 405–419.

Fuerst, F., McAllister, P., Nanda, A. and Wyatt, P. (2013) An investigation of the effect of EPC ratings on house prices. Department of Energy and Climate Change, London.

Galvin, R. (2011) Discourse and materiality in environmental policy: The case of German Federal policy on thermal renovation of existing homes. PhD thesis, University of East Anglia.

Hargreaves, T., Nye, M. and Burgess, J. (2010) Making energy visible: A qualitative field study of how householders interact with feedback from smart energy monitors. *Energy Policy*, 38, 6111–6119.

Harris, J., Anderson, J. and Shafron, W. (2000) Investment in energy efficiency: A survey of Australian firms. *Energy Policy*, 28(12), 867–876.

HM Government (2010) *Building Regulations Part L1b Conservation of Fuel and Power in Existing Buildings*, HMSO, London.

IEA (2007) Mind the gap: Quantifying principal–agent problems in energy efficiency. Organisation for Economic Development/International Energy Agency, Paris.

Institute for Sustainability (2012) Retrofit insights: Perspectives for an emerging industry. Key findings: analysis of a selection of Retrofit for the Future projects. www.instituteforsustainability.co.uk/uploads/File/KeyFindings.pdf, accessed 28 May 2015.

Jankel, Z. (2013) Delivering and funding housing retrofit: A review of community models. ARUP, Institute for Sustainability, London.

Jones, C. (2012) An analysis of payback times and importance of embodied carbon in retrofit of domestic houses. BSc dissertation, University of East Anglia.

Jones, P., Lannon, S. and Patterson, J. (2013) Retrofitting existing housing: How far, how much? *Building Research and Information*, 41(5), 532–550.

Juan, Y.-K., Gao, P. and Wang, J. (2009) A hybrid decision support system for sustainable office building renovation and energy performance improvement. *Energy and Buildings*, 42, 290–297.

Kok, N., Eichholtz, P., Bauer, R. and Peneda, P. (2010) Environmental performance: A global perspective on commercial real estate. The European Centre for Corporate Engagement, Maastricht University School of Business and Economics, Maastricht.

Lapillonne, B., Pollier, K. and Sebi, C. (2012) Energy efficiency trends in buildings in the EU. Energdata, Grenoble.

Lewis, J. and Smith, L. (2014) Breaking barriers: An industry review of the barriers to whole house energy efficiency retrofit and the creation of an industry action plan. Energy Efficiency Partnership for Buildings and the National Energy Foundation, Milton Keynes.

Ma, Z., Cooper, P., Daly, D. and Ledo, L. (2012) Existing building retrofits: Methodology and state-of-the-art. *Energy and Buildings*, 55, 889–902.

Maller, C.J. and Horne, R.E. (2011) Living lightly: How does climate change feature in residential home improvements and what are the implications for policy? *Urban Policy and Research*, 29(1), 59–72.

MOHURD (2007) Interim management measures of heat metering and energy efficiency retrofit funds for existing residential buildings in northern heating regions of China. Department of Economic Development of MOF, Washington and Beijing.

National Refurbishment Centre (2012) Refurbishing the nation: Gathering the evidence. www.rethinking refurbishment.com/filelibrary/nrc_pdf/NRC_reportSEP2012web.pdf%20, accessed 28 May 2015.

NHBC Foundation (2013) Assessment of MVHR systems and air quality in zero carbon homes. www.nhbcfoundation.org/Publications/Primary-Research/Assessment-of-MVHR-systems-and-air-quality-in-zero-carbon-homes-NF52, accessed 28 May 2015.

Odyssee (2009) Database for energy efficiency indicators in Europe. www.odyssee-indicators.org, accessed 5 June 2015.

Owens, S. and Driffill, L. (2008) How to change attitudes and behaviours in the context of energy. *Energy Policy*, 36(12), 4412–4418.

Pacific Northwest National Laboratory (2011) A guide to energy audits. www.pnnl.gov/main/publications/external/technical_reports/pnnl-20956.pdf, accessed 28 May 2015.

Palmer, J. and Cooper, I. (2013) United Kingdom housing energy fact file. Department of Energy and Climate Change, London.

Power, A. (2008) Does demolition or refurbishment of old and inefficient homes help to increase our environmental, social and economic viability? *Energy Policy*, 36(12), 4487–4501.

Ravetz, J. (2008) State of the stock: What do we know about existing buildings and their future prospects? *Energy Policy*, 36(12), 4462–4470.

Rosenow, J. and Eyre, N. (2012) The Green Deal and the Energy Company Obligation: will it work? *9th BIEE Academic Conference – European Energy in a Challenging World: The Impact of Emerging Markets*. St John's College, Oxford, 19–20 September.

Rosenow, J. and Galvin, R. (2013) Evaluating the evaluations: Evidence from energy efficiency programmes in Germany and the UK. *Energy and Buildings*, 62, 450–458.

Roth, K.W., Westphalen, D., Feng, M.Y., Llana, P. and Quartararo, L. (2005) Energy impact of commercial building controls and performance diagnostics: Market characterization, energy impact of building faults and energy savings potential, final report. TIAX LCC, Lexington.

Royal Academy of Engineering (2010) Engineering a low carbon built environment: The discipline of building engineering physics. www.raeng.org.uk/publications/reports/engineering-a-low-carbon-built-environment, accessed 28 May 2015.

Royston, S., Daly, M. and Foulds, C. (2014) Know-how, practices and sustainability. In: Foulds, C. and Jensen, C.L. (eds), *Practices, the Built Environment and Sustainability: A Thinking Note Collection*, GSI, DIST, BSA CCSG, Cambridge, Copenhagen, London.

Shilei, L. and Yong, W. (2009) Target-orientated obstacle analysis by PESTEL modelling of energy efficiency retrofit for existing residential buildings in China's northern heating region. *Energy Policy*, 7(6), 2098–2101.

Shove, E. (2003) *Comfort, Cleanliness and Convenience: The Social Organization of Normality*, Oxford, Berg.

State of Wisconsin (2014) Energy assistance and weatherization assistance programs. Legislative Audit Bureau, Department of Administration. http://legis.wisconsin.gov/lab/reports/14-8full.pdf, accessed 5 June 2015.

Šumarac, D. (2010) Energy efficiency of buildings in Serbia: State-of-the-art and perspectives. *Termotehnika*, 36(1), 11–29.

Sustainable Development Commission (2006) Stock take: Delivering improvements in existing housing. www.sd-commission.org.uk/data/files/publications/Stock_Take.pdf, accessed 28 May 2015.

Sweett (2014) Retrofit for the Future: Analysis of cost data. https://retrofit.innovateuk.org/documents/1524978/1866952/Retrofit%20for%20the%20Future%20-%20analysis%20of%20cost%20data%20report%202014, accessed 28 May 2015.

Todorović, M.S. (2010) National energy efficiency action plan of buildings in Serbia: An approach to the large scale municipal energy refurbishment. *REHVA Journal*, 47(6), 22–26.

TSB (2013a) Retrofit revealed: The Retrofit for the Future projects – data analysis report. Technology Strategy Board, Swindon.

TSB (2013b) Low energy building database. Technology Strategy Board, http://retrofitforthefuture.org, accessed 5 June 2015.

UK Commission for Employment and Skills (2014) Employers skills survey UK results 2013. www.gov.uk/government/uploads/system/uploads/attachment_data/file/327492/evidence-report-81-ukces-employer-skills-survey-13-full-report-final.pdf, accessed 28 May 2015.

USBED (2009) US building energy data book. http://buildingsdatabook.eren.doe.gov/docs/DataBooks/2009_BEDB_Updated.pdf, accessed 5 June 2015.

Westminster Sustainable Business Forum and Carbon Connect (2013) Energy demand: reducing energy demand in the commercial sector. www.policyconnect.org.uk/wsbf/research/building-efficiency-reducing-energy-demand-commercial-sector, accessed 28 May 2015.

Wilson, C., Chryssochoidis, G. and Pettifor, H. (2013) Understanding homeowners' renovation decisions: Findings of the VERD project. UK Energy Research Centre, London.

Yan, D., Zhe, T., Yong, W. and Neng, Z. (2011) Achievements and suggestions of heat metering and energy efficiency retrofit for existing residential buildings in northern heating regions of China. *Energy Policy*, 39(9), 4675–4682.

Zhu, Y. (2013) Energy and economic analysis of renovation measures for rural residential buildings in Northern China. MPhil in Environmental Design in Architecture, Department of Architecture, University of Cambridge.

6 The Passivhaus energy-efficiency standard

6.1 Introduction

The previous chapters have identified the importance of reducing the energy demand of our building stock, with energy-efficiency codes and standards (be they voluntary or mandatory) commonly talked about as a means to do this. We introduce the Passivhaus standard in this chapter, which is a voluntary energy-efficiency standard for residential and non-residential buildings that follows a set of specific performance criteria established by the Passivhaus Institute in Germany. This chapter explores what constructing to the Passivhaus energy-efficiency standard entails; from its basic foundations regarding technology and history, to exploring the implications of introducing these technologies to the occupants themselves, as well as those who govern that process, including landlords and the broader industry. It is important to consider such issues because the Passivhaus standard is widely promoted, particularly in industry and policy circles, as an exemplary means of lowering residential energy demand while delivering comfort.

It is important to emphasise that we use the term 'Passivhaus', as opposed to the English translation, Passive House or (the less commonly referred, but technically correct) Passive Building. This is because Passivhaus is an approach in itself, which can only be achieved through meeting several very specific criteria that have been approved by the Passivhaus Institute. To call it Passive House, as many do, we see as potentially confusing. For example, Passivhaus is certainly not passively ventilated, because of its reliance on an active ventilation system. Moreover, as Schiano-Phan *et al.* (2008, p. 2) discuss, 'passive house generally means any house constructed in line with the principles of passive solar design' in Southern Europe (e.g. Spain, Italy, Portugal, Greece).

6.2 Introducing Passivhaus

6.2.1 What is the Passivhaus standard?

Passivhaus is a voluntary standard for building energy efficiency, although it has been made compulsory in some districts and municipalities of Europe. The standard originated

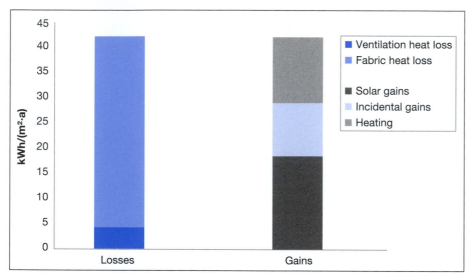

Figure 6.1 Heat energy losses and gains in a typical Passivhaus home (kWh/(m²·a)).
Source: Feist *et al.* (2005, p. 1187).

in Germany in the 1990s, but is being utilised internationally. Although mainly applied to new buildings, a variant, 'EnerPHit', is also available for retrofitting. The Passivhaus standard is not solely intended for the residential sector; commercial buildings such as offices, schools and shops are also being constructed to this standard. To officially meet the Passivhaus standard, a building must be officially certified by the Passivhaus Institute (PHI) based in Darmstadt, Germany, or by 'Approved Passivhaus certifiers' in different countries. However, due to cost implications, this certification is not always sought.

Passivhaus buildings achieve most of their energy reductions through a 'fabric first' approach that minimises the need for space heating and cooling. Buildings are so well insulated, with minimum ventilation heat loss, that they stay warm by using the heat from people, solar gain and appliances (incidental gains) (Figure 6.1). For example, in theory one could significantly help to heat their home by using their hairdryer or vacuum cleaner.

It is this approach that makes Passivhaus different from buildings in the past that claimed to be passively heated, which relied almost solely on south-facing solar gain. It is also markedly different to the many current low-energy buildings that often rely heavily on renewables. It is anticipated by many that Passivhaus (at the very least, inspired) design will become increasingly standard practice within industry in the next ten years, as we seek to reduce the energy demand of our built environment.

6.2.2 *Common design characteristics*

This section explores the fundamental design principles of Passivhaus, while Section 6.2.3 focuses on the specific Passivhaus assessment criteria. As such, we now clarify what constructing to Passivhaus tangibly means for the design and construction of a development (i.e. what are its salient features).

Its common design characteristics are as follows.

Super insulation

Insulation plays an important role in reducing fabric heat losses and, as such, it is common for additional wall and floor insulation and triple glazing to be installed, in addition to loft insulation being 500 mm thick (assuming standard Rockwool insulation is used). There is no one specific way to insulate a Passivhaus building because it depends on the broader construction approach taken (i.e. how the assessment criteria in Section 6.2.3 are specifically adhered to). For more details on the thermal performance of various insulation materials, see Section 2.4.3.

Minimising thermal bridges

This super insulation needs to wrap continuously around the building, and in doing so leave no gaps – for instance, where the walls join the floor and roof, so as to minimise thermal bridges. A thermal bridge is an area of the building envelope where the heat flow is different to its adjacent areas (see Section 2.4.4 for more details). Thermal bridges occur where materials that are good conductors of heat (e.g. steel) interrupt the insulation layer even in small areas. A commonly described analogy for this is that of the tea pot (the building) and tea cosy (insulation). Any sort of 'hole' (thermal bridge) in the tea cosy will provide a means for the heat to flow out.

Airtightness and mechanical ventilation and heat recovery (MVHR)

By airtightness, we mean how resistant a building envelope is to inward or outward (unintentional) air leakage. In meeting Passivhaus' airtightness requirements, and thereby restricting natural airflow through a building, it is extremely difficult to then not have to install mechanical ventilation for air quality (and thereby health) purposes. Further, a heat recovery element is essential because otherwise the ventilation system will not be as energy efficient as it could be, which could mean that airtightness does not end up delivering the necessary (and anticipated) reductions in net energy demand. Therefore, MVHR systems are each equipped with heat exchangers, which use the heat of the outgoing warm air to heat the stream of incoming cool air (see Chapter 7 for more details). Typically, 80–90 per cent of the heat in the outgoing air is transferred to the supply air. Without this, potentially large amounts of heat could be lost as a by-product of adequately ventilating the building (for more on ventilation, see Chapter 7).

Passive solar gain

To optimise the use of direct solar gains, and thus reduce the amount of energy used by the heating system, Passivhaus buildings should ideally have large areas of south-facing (assuming a Northern Hemisphere location) glazing that account for at least 40 per cent of the elevation's surface area. To accompany these, much smaller north-facing windows are typical because there is little direct solar gain. It is important to note that since glazing has a higher fabric heat loss rate than well-insulated external walls, larger north-facing windows will lose more heat energy than they would provide. This often therefore results in toilets, bathrooms, cloakrooms and stairs being located on the north side, as they have different daylighting needs. Moreover, such is the effectiveness of being south-facing, that solar shading (e.g. brise soleil) is commonly installed to inhibit overheating in summer. Nevertheless, while facing south (or rather, within 30 degrees from south) is the ideal scenario, buildings can still be Passivhaus-certified if east-/west-facing. It merely requires a slight change in design. Site-related restrictions for new builds and, in particular, retrofitting projects may not provide the luxury of facing south.

Energy-efficient electrical appliances

To meet the low primary energy demand, energy efficient appliances are needed. The benefit of this is that efficient appliances use relatively less energy, generating less heat inside the building, and thereby inhibiting overheating. Ideally, energy-efficient appliances should be pre-installed and sold with the building itself (e.g. as part of a fitted kitchen), so as to minimise the chance of residents bringing in their own, perhaps inefficient, appliances from their previous home. The provision of appliances is more likely for homes that are being sold because, for instance, landlords may not be keen to take on the maintenance responsibility for appliances used by their renters.

Complementary renewable technologies

As energy efficiency measures can only go so far in achieving the required (low) primary energy demand, designers are increasingly being drawn towards the consideration of renewable energy resources. An example of an especially complementary renewable would be solar thermal energy, because Passivhaus' enhanced airtightness removes the need for heating system temperatures higher than 50 °C (Badescu and Sicre, 2003a; 2003b). Therefore, while renewables are not a requirement of Passivhaus certification, a new standard – 'Passivhaus plus' – was recently introduced to include renewables as part of certification. It was first introduced by the PHI in 2014 to make Passivhaus compliant with the EU carbon-reduction targets.

The Passivhaus standard is not a 'bolt-on' feature or additional attachment to design. Instead it represents an integrated design, construction and commissioning process that more holistically considers a building's architecture, structure, and mechanical and electrical systems. It is central to how the building functions and is ultimately used. Therefore designs cannot be easily changed once work begins, making accurate designing even more paramount than normal. Such is the importance of detail, and that construction is both 'tight' and 'right', that it is common for building plans to be at a significantly larger scale than normal, so as to provide additional clarity when interpreting these plans when on-site.

6.2.3   Requirements of the Passivhaus standard

Unlike the UK government's Code for Sustainable Homes (CfSH) and Standard Assessment Procedure (SAP), which state an explicit carbon dioxide target, meeting the Passivhaus standard requires adherence to certain minimum energy-efficiency requirements, as set by the PHI. While these requirements do tend to pull designers towards certain common characteristics (as identified in Section 6.2.2), they do not dictate a particular method of construction, as indicated by the following Passivhaus building projects that vary in their construction approach:

- Ravensburg Kunsthalle, Germany: concrete structure with a cavity wall, brick-facing.
- Denby Dale, Yorkshire, UK: traditional cavity wall construction adapted to meet Passivhaus requirements.
- Stories Mews, Dulwich, UK: structurally insulated panel (SIP) construction.

There is little flexibility in meeting all of the Passivhaus requirements, although each Passivhaus project does have the freedom to choose between meeting either an annual space heating/cooling demand target or a peak heating load target. Such a choice has been allowed because both options still ensure highly efficient heating performance.

Thus, all of the following four requirements must be met:

1a Space heating/cooling demand must be no higher than 15 kWh/(m²·a).

As with all these requirements, the standards do not change depending on geography or context. Therefore, this energy demand (i.e. energy used to regulate indoor temperatures, humidity and air quality) must be adhered to whether, for instance, the Passivhaus building is in a tropical or colder climate. Indeed, this is why both heating and cooling are explicitly included, although we note that criteria for cooling is actually slightly more complex than just reaching the 15 kWh/(m²·a) target.

Instead of 1a, adhering to 1b is an acceptable alternative.

1b Space heating/cooling load must be no higher than 10 W/m²

The specific heating load is essentially the peak power of heating the building at one given moment in time (not over a period of time, as 1a addresses). In energy terms, 10 W/m² is approximately equivalent to two 60 W light bulbs in a 12 m² room.

2 Annual primary energy demand must be no higher than 120 kWh/(m²·a)

This includes all building applications, unlike UK Buildings Regulations for instance, and thereby covers all the energy demands of the building, including appliances. Note that there is an explicit focus on primary energy, rather than delivered energy. This target also ensures that 1a or 1b cannot be achieved by increasing the energy demand of (and thereby utilising the passive heating from) other end-uses, such as electricity for appliances.

3 Airtightness must be less than 0.6 ach, at a pressure difference of 50 Pa

Accidental leakage through the fabric of the building must be reduced to less than 0.6 air changes per hour (ach). Increasing airtightness to such an extent will ensure that minimal amounts of heat are lost through draughts, which (if the case) would be detrimental to achieving the heating demand/load targets of 1a and 1b. Since 0 ach is not targeted and indeed could never be achieved, describing Passivhaus buildings as 'airtight', as many do, is technically incorrect. It is just that compared to conventional buildings, Passivhaus buildings are extremely airtight. It is also worth emphasising that certified Passivhaus buildings will, whatever their size and specification, have undergone mandatory airtightness tests, also often referred to as 'blower door' tests.

4 *Frequency of overheating – must not be more than 25 °C for more than 10 per cent of the year*

Passive heat gains (solar and occupant-related) are so effective at reducing the heating load that efforts are needed to ensure that overheating does not occur and thermal comfort is maintained. Thus, this requirement stipulates that the indoor temperatures of Passivhaus-certified buildings cannot go above 25 °C for 10 per cent or more of the year. Note that this assessment of overheating is not done post-occupancy, but is based upon modelled assumptions during the design process.

6.2.4 *History of Passivhaus*

The underlying concepts of Passivhaus were developed by Wolfgang Feist (at the time: Institute for Housing and the Environment, in Darmstadt, Germany) and Bo Adamson (Lund University, Sweden), both of whom had worked and published extensively on low-energy building design. When reflecting upon the early developmental stages of the Passivhaus standard, Feist offered the following reflections:

> I was working as a physicist. I read that the construction industry had experimented with adding insulation to new buildings and that energy consumption had failed to reduce. This offended me – it was counter to the basic laws of physics. I knew that they must be doing something wrong. So I made it my mission to find out what, and to establish what was needed to do it right.
>
> (Feist, in Reason and Clarke, 2008, p. 2)

In 1990, Feist and Adamson's early Passivhaus concepts were applied to a new-build development in Darmstadt-Kranichstein, with residents occupying the four terrace houses in 1991. Energy demand was measured and, relative to the wider German housing stock, savings in the region of 85–90 per cent were consistently achieved year after year. Furthermore, a space heating demand of as low as 10 kWh/(m²·a) was achieved.

The International Passive House Association (iPHA) accredits Amory Lovins, a leading US energy-efficiency figure, with the role of moving Passivhaus from this demonstration phase to a position whereby energy savings could be targeted on a wider scale. The iPHA tells the story that Amory Lovins visited the first Passivhaus (Darmstadt) project in 1995 and is quoted as saying:

> No, this is not just a scientific experiment. This is the solution. You will just have to redesign the details in order to reduce the additional costs – and that will be possible, I am convinced.
>
> (Lovins, in International Passive House Association, 2013)

Following a second development (in Groß-Umstadt in 1995), it was this sort of encouragement, in tandem with the Darmstadt development's proven energy savings, which led to Feist translating the design approach into the Passivhaus standard. The Passivhaus Institute was founded in September 1996 to both further the energy-efficient Passivhaus building standard and act as a body for certification.

Following the official founding of both the Passivhaus standard and the Passivhaus Institute, expansion occurred, albeit relatively gradually. For example, Austria's first Passivhaus homes were built in 2000, with Sweden's being in 2001, Italy's in 2002, the USA's in 2003, Ireland's in 2005 and the UK's in 2009 (Cox, 2005; Passivhaus Trust, 2012a). For further details on the recent growth and current number of Passivhaus buildings internationally (majority in Germany and Austria), see Section 6.3.

In addition to international growth, the standard has now become widely used across different sectors and various building types. Indeed, this is nicely demonstrated by the international six contrasting award categories in the '2014 Passive House Awards': single-family homes; apartment buildings; educational buildings; office and special-use buildings; retrofits; and regions (Passive House Institute, 2014c).

6.2.5 The EnerPHit standard

The success of the Passivhaus standard, which fundamentally targets energy-efficiency improvements for new builds, combined with the increasing need to improve the energy efficiency of our existing building stock (see Chapter 5), has led to an equivalent Passivhaus retrofitting standard being introduced: the EnerPHit standard. After being piloted in 2010 and early 2011, the Passivhaus Institute launched the EnerPHit standard at their annual conference in May 2011.

Constructing new builds poses fewer constraints and more opportunities for improved levels of energy efficiency than retrofitting, which is typically more challenging because of the existing structural context. For instance, it is more likely for retrofitted buildings to have thermal bridges simply because the existing fabric is already in situ, hence efforts are usually made to minimise, rather than extensively design out, thermal bridges. Indeed, although it is possible (e.g. Grove Cottage Passivhaus, Hereford, UK), retrofitting solid-walled Victorian properties is particularly challenging due to their complex detailing.

EnerPHit assessment criteria inherently acknowledge such difficulties and, as such, are more flexible in their approach compared to the Passivhaus standard. This flexibility, however, makes the certification more complicated because there are more criteria to be assessed.

The key assessment criteria are as follows (Feist, 2013):

- Annual specific space heating demand must be less than 25 kWh/(m²·a).
- Airtightness must be less than 0.6 ach (known as the 'target value'), but less than 1.0 ach may be accepted if supporting evidence is provided (known as 'limit value').
- Evidence to demonstrate how issues of internal moisture and humidity have been addressed.
- Internal temperatures must be less than 25 °C for no more than 10 per cent of the year.
- Evidence to demonstrate that areas near windows do not have significantly lower temperatures than the rest of the building fabric (e.g. through triple glazing or equivalents, or additional heaters).
- Evidence to demonstrate that it is an existing building and not a new build.

If the 25 kWh/(m²·a) space heating demand maximum is not met, EnerPHit certification is still possible through meeting the following criteria, which largely target specific building fabric elements (Feist, 2013):

- External walls: more than 75 per cent of wall area must be externally or internally insulated, with U-values that are no higher than 0.15 W/(m²·K). Should there be evidence that external insulation is impracticable, then internally insulated external walls with slightly higher U-values may be permitted.
- Roofs (and top-floor ceilings): U-values that are no higher than 0.12 W/(m²·K).
- Floors: U-values that are no higher than 0.15 W/(m²·K).
- Windows: U-values that are no higher than 0.85 W/(m²·K).
- External doors: U-values that are no higher than 0.80 W/(m²·K).
- Thermal bridges: these cannot always be eliminated (without an unreasonable amount of effort), particularly in comparison to new builds, but evidence is needed to demonstrate that they have been avoided or minimised as much as possible.
- Mechanical ventilation: systems must have a heat recovery efficiency of at least 75 per cent and an electrical efficiency of at least 0.45 Wh/m³.

6.2.6 Becoming Passivhaus certified

The assessment criteria presented in Section 6.2.3 (Passivhaus) are the basis for Passivhaus certification (likewise for Section 6.2.5 and EnerPHit certification). Similar to other building standards, in assessing whether such criteria are met, actual measured post-occupancy data are not used. Indeed, the certification process ensures that the construction has abided by the design commitments, of which the modelling results of the Passivhaus Planning Package (PHPP) (an Excel-based design tool produced by the Passivhaus Institute) forms a considerable part. The latest update (version 8.5) to the PHPP was in April 2014. Since 2013, the PHPP has included algorithms that account for various climatic regions around the world, thereby allowing building projects to apply for Passivhaus certification, whatever their location (Passive House Institute, 2014a).

The certification process specifically includes a:

- journal recording construction details on-site (including relevant photographs);
- thermal bridge analysis, using specialist software (e.g. Therm);
- compression/decompression test to verify airtightness levels;
- the submission of the relevant PHPP modelling.

In understanding the certification process, it is important to distinguish between the following roles:

- Certified Passivhaus Designers: these are accredited with the required industry standard to work professionally as Passivhaus designers across the world. Accreditation is usually achieved through participating in a PHI-recognised course that focuses on the principles and methodology required for Passivhaus design. The Certified Passivhaus Designer status can also be achieved through successfully designing and constructing a Passivhaus-certified building (although this is less common, as it is regarded to be a more difficult route). Consultants often adopt the Certified Passivhaus Designer role, through helping guide the project architect (who perhaps may not actually be certified) through the technical design process and modelling.
- Approved Passivhaus Certifiers: these have been accredited by the PHI to, on its behalf, assess and issue a quality-assured version of the Passivhaus certificate, which would have in turn been provided by the Certified Passivhaus Designers.

As of April 2013, there are over 4,000 Certified Passivhaus Designers in the world (Passive House Institute, 2014b). There are considerably fewer Approved Passivhaus Certifiers. For example, in the UK there are currently only four approved certifiers: (1) Building Research Establishment; (2) Cocreate Consulting; (3) Mead: Energy and Architectural Design; and (4) Warm Associates.

6.3 Uptake of the Passivhaus standard

6.3.1 World-wide focus

In 2012 there were around 50,000 Passivhaus buildings in the world (International Passive House Association, 2014a). Since the Passivhaus assessment criteria do not stipulate a particular design and/or construction method, nor particular climatic requirements, there are buildings certified to the Passivhaus standard all over the world. Indeed, by 2014, Passivhaus buildings have been constructed in 35 different countries, including Australia; Canada; China; Finland; Ireland; Luxembourg; Mexico; Russia; South Korea; and Spain, to name only a few. Nevertheless, despite this wide-ranging reach, most of the world's Passivhaus buildings have been constructed in Europe, particularly in Germany and Austria.

Around 25 million inhabitants live across, as the International Passive House Association call them, 27 EU Passivhaus 'hot spots' (Mekjian, 2011). These hot spots have Passivhaus embedded in the specific area's planning policies and/or building regulations as a minimum requirement. The majority of these hot spots are in Germany; Table 6.1 details their specific legislative requirements. The two other significant hot spots are in Belgium. First, in June 2013 the province of Antwerp became committed to all new buildings and complete renovations in the public sector being Passivhaus. Second is the capital region of Brussels, which goes even further by demanding that all new buildings and retrofits, whatever the sector, will have to be Passivhaus from January 2015 onwards (International Passive House Association, 2014b). Therefore policy-makers are beginning to publically commit legislation to the Passivhaus brand, making it a clear contributor to Europe's future dwelling stock.

While in 2008 there was an (albeit failed) European Parliament Resolution proposed that all new EU buildings must reach the Passivhaus (or equivalent) standard (European Parliament resolution of 31 January 2008 on an Action Plan for Energy Efficiency: Realising the Potential, 2007/2106 INI), there have been few other cross-national legislative attempts to make Passivhaus mandatory. Other countries have been significantly influenced by the Passivhaus standard, or at least it has triggered a debate around constructing super-insulated and airtight buildings. The building regulations and planning guidelines of these countries may not require the Passivhaus brand, as per the Passive House Institute in Darmstadt, be met as a minimum. For instance, from 2015, Finnish dwellings must meet Finland's own 'Passive House' definition (again, emphasising the need for differentiation from the Passivhaus brand), which is similarly based on achieving high energy reductions through super-insulation and airtightness (GBPN, 2013). Likewise, the MINERGIE-P standard in Switzerland is also very similar to the Passivhaus standard. Atanasiu *et al.* (2011) show how national building policies across Europe, and specifically how these policies define a zero-energy house, have been inspired by Passivhaus.

The Passivhaus standard is thus influencing energy and building policies across Europe, which are in turn beginning to attract further attention globally. It is internationally regarded

Table 6.1 Passivhaus legislation: looking closer at German states, cities and districts

Location	Description
States	
Bavaria	All public new builds will be built to Passivhaus standard. Agreed by council ministers on 19 July 2011.
Hesse	All public new builds, as of September 2012, must have an energy performance equivalent to the Passivhaus levels.
Rhineland-Palatinate	All public new build and retrofit projects must be reviewed to determine whether the buildings can feasibly meet the Passivhaus standard. This was from 2010 onwards, in response to the target of becoming a carbon-neutral state administration.
Saarland	All new public buildings must be Passivhaus, in addition to it being a central guideline for the retrofitting of all public buildings.
Towns and cities	
Bremen	In response to the city's target of 50 per cent lower CO_2 emissions from public buildings, all new public buildings will be built to Passivhaus. Agreed on 25 August 2009, came into force on 1 January 2010.
Cologne	As of 26 April 2008, all buildings built in the city must be Passivhaus.
Frankfurt	All new public buildings, in addition to any other buildings built as part of the public–private partnership model, must meet the Passivhaus standard. Originated from a 6 September 2007 resolution.
Freiburg	All new homes had to meet the Passivhaus standard from 2011 onwards. Originated from a 22 July 2008 resolution.
Hamburg	Municipal funding for new housing was, from 2012 onwards, only available to projects meeting the Passivhaus standard.
Heidelberg	All new public buildings to meet Passivhaus criteria. Also, those buying land from the municipality for new-build construction will be obliged to adhere to the Passivhaus standard (unless economically unfeasible).
Kempton	All new public buildings must meet Passivhaus requirements, unless technically or economically unrealistic.
Leipzig	All new public buildings and buildings built by the municipality and through public–private partnerships must be Passivhaus. Originated from a 19 March 2008 resolution.
Leverkusen	As of 16 February 2009, all new buildings must be Passivhaus. In addition, a target of retrofitting 50 per cent of existing buildings to Passivhaus standard was also put in place.
Nuremberg	All new buildings must be Passivhaus.
Walldorf	All new buildings built for and by the municipality must be Passivhaus. Originally passed on 20 July 2010.
Districts	
Lippe	All new public buildings will be planned and constructed to the Passivhaus standard. Passivhaus components will also be used when retrofitting the authority's existing building stock.
Darmstadt-Dieburg	All new schools, managed by Da Di-Werk, are to be Passivhaus.

Source: based on data collated by the International Passive House Association (2014b); see this source for links to each respective legislation document.

as a solution to reducing energy demand in the built environment. It is because of the momentum the Passivhaus standard is gathering that many buildings are being constructed in accordance with the standard, but are not being officially certified. A consequence of this is that we can only ever know the number of Passivhaus-certified buildings, which would exclude many buildings that most likely still label themselves as Passivhaus.

6.3.2 Passivhaus in the UK

The first Passivhaus-certified building in the UK was a multi-purpose office building in Machynlleth, Wales that was completed in August 2008, and occupied in January 2009. Table 6.2 provides background information on the first new-build Passivhaus dwelling, office building and educational building in the UK.

Many project teams (and their clients) are deciding not to officially certify their units, primarily because of the cost associated with certification, in addition to Passivhaus construction already being more expensive. Instead, designers and constructers may have to prove in other ways (to the client) that the standard has been reached, without having it officially rubber-stamped by the Institute itself. For example, the project in Tigh-Na-Cladach, Scotland, only had one of its ten units officially certified (Passivhaus Trust, 2013d) (although it is possible for one Passivhaus certificate to relate to a set of buildings). It is important to emphasise that, whether in the UK or in other countries, numbers of Passivhaus buildings tends to translate into the total Passivhaus-*certified* buildings. Thus, it is more difficult to gauge how many households and/or sets of commercial occupants may have been exposed to Passivhaus-related technological standards.

The number of Passivhaus-certified new-build construction projects being completed year on year is increasing (Figure 6.2). Nevertheless, as the total number of Passivhaus units indicates, while Passivhaus is gathering momentum it still remains a relatively niche building standard in the UK. Indeed, such is the state of knowledge in the UK Passivhaus

Table 6.2 Background information on the first Passivhaus-certified new builds in the UK

First UK Passivhaus-certified . . .	Location	Completion date	Project	Source and for more information
Dwelling	Wales, and the UK	October 2009	Y Foel, Machynlleth	Tiramani (2013); Passivhaus Trust (2013c)
	England	August 2010	Underhill House, Moreton-in-Marsh	Passivhaus Trust (2013e)
	Scotland	April 2010	Tigh-Na-Cladach, Dunoon	Ford and Hill (2011)
Community/ office building	Wales, and the UK	August 2008	Canolfan Hyddgen, Machynlleth	Passivhaus Trust (2013a)
	England	February 2010	Centre for Disability Studies, Essex	Simmonds Mills (2010)
Educational building	England, and the UK	February 2010	Hadlow College, Kent	Passivhaus Trust (2012b)

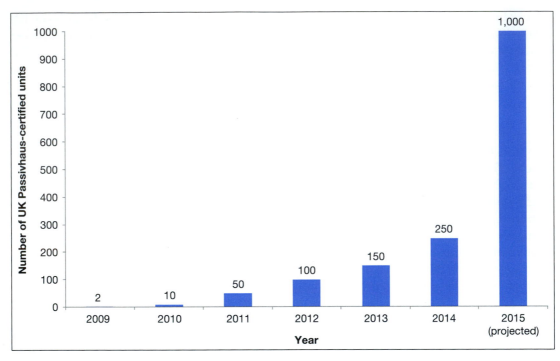

Figure 6.2 Cumulative growth in a number of UK Passivhaus-certified projects (2009–2015).
Source: Bootland (2014).

community that frequent attempts are made to learn from German counterparts (despite it still not being 'mainstream' in Germany either), as German industry is much more familiar with the Passivhaus standard (e.g. Cutland, 2012). One could therefore speculate that the UK public is not only likely to be unfamiliar with Passivhaus technologies, but also unaware of Passivhaus as a concept more generally.

It is anticipated by many that Passivhaus will increasingly become standard practice over the next decade (Feist in McCabe, 2012; Boardman, 2012). Mainstream UK media have also reported on the Passivhaus standard in recent years (McGhie, 2008; Anderson, 2013; Earley, 2013, Krestovnikoff and Poyntz-Roberts, 2013), with one headline asking whether Passivhaus is 'the housing standard of the future?' (Jenkinson, 2010). Such articles have predominantly been complimentary to the standard. Media interest was perhaps initially sparked by UK Channel 4's popular *Grand Designs* television programme, which had an episode each on the design and construction of the Underhill House Passivhaus project (Tebbutt, 2010) and the Crossway House Passivhaus project (Hawkes Architecture, 2014). Looking towards prominent politicians, Chris Huhne, during his period in office as UK Secretary of State for Energy and Climate, stated in October 2010 that he 'would like to see every new home in the UK reach the (Passivhaus) standard' (DECC, 2010). However, these words of support have never materialised in a UK policy that advocates Passivhaus.

Despite calls from the Passivhaus community, the UK has no current plans to make the Passivhaus standard mandatory for all new buildings, whether on a national or area-specific basis as Germany has done (Table 6.1). It is unlikely that this will happen, in the short-term at least, in part because the energy standards of the UK's Building Regulations, Code for Sustainable Homes initiative (now voluntary, but did previously direct building regulations) and Standard Assessment Procedure (SAP) software have explicit carbon dioxide targets, whereas Passivhaus focuses on energy efficiency. This inherent difference could lead to low carbon supplies being prioritised over energy efficiency. Saying this, the work from the UK's Zero Carbon Hub on the (Passivhaus-style) Fabric Energy Efficiency Standard, and the adoption of fabric energy efficiency into the UK government's Building Regulations in 2013, could suggest that energy efficiency is gaining more traction.

With reference back to the UK's SAP, it should additionally be noted that past analyses (albeit on slightly older versions of the software) have argued that SAP has numerous underlying assumptions that make Passivhaus design less attractive (Reason and Clarke, 2008). SAP was developed in the 1980s from studying dwellings with poor insulation and high heat loss, hence it rewards buildings that have renewables yet are 'leaky' and thus have relatively high heat loss (i.e. enables prioritisation of energy supply over energy demand in decarbonisation). Unsurprisingly, the Passivhaus Planning Package (PHPP), being Passivhaus' design and certification tool, thus rewards designs that prioritise the reduction of energy demand (Reason and Clarke, 2008).

Unless there is a shift in UK Building Regulations (e.g. to consider requiring Passivhaus certification) and/or how they are assessed (e.g. moving away from the dated SAP), we would speculate that the likelihood of Passivhaus becoming more mainstream in the near future remains low. Therefore, adhering to the Passivhaus standard will remain voluntary in the UK, and thereby will most likely continue to rely on designers, constructers and their clients having confidence that the Passivhaus standard will save energy and/or provide occupants with a comfortable living environment. For example, Hastoe Housing Association has made a commitment for at least 20 per cent of its new developments to be Passivhaus. However, even if Passivhaus is never actually made mandatory, it could be argued that if it helps to inspire improvements to building standards more generally, then the difference between current mainstream and novel Passivhaus forms of construction would become less and less significant.

6.4 Implications of targeting the Passivhaus standard

6.4.1 Energy performance of Passivhaus buildings

The primary focus of Passivhaus design is a significant reduction in the space heating demand, because in the countries of origin this is the dominant use of energy in the built environment. However, the standard also recognises the need to reduce total energy consumption, and to take account of the heat gains arising from the use of appliances and hot water distribution. Indeed, the PHPP manual states that 'highly efficient use of household electricity is essential' (Passive House Institute, 2013, p. 16). The design should thus ideally incorporate 100 per cent low-energy lighting, and the developer should supply energy-efficient appliances. The 'electricity' part of the design will not only assume such appliances, but will also assume that they are used as little as practicable; that is, that the residents are motivated to conserve energy. In the current version of PHPP, the designer is encouraged to be realistic in accounting for electricity use from the expected number of occupants, rather than the number calculated by PHPP (which is likely to be a smaller number because of the generous, by UK standards, space allowance per person). This choice of 'standard' or 'user-determined' must be agreed with the certifier. Early UK Passivhaus designs tended to use the standard figure, which made it easier to meet the primary energy target. However, landlords were unable to provide appliances, and tenants brought existing items and practices with them, resulting in little change in electricity consumption from their previous homes.

The consequence of this is that the Passivhaus standard has struggled to reduce electricity demand. In fact, electricity demand commonly increases slightly (when moving from a non-Passivhaus to Passivhaus building) due to the occupants using roughly the same electrical appliances in roughly the same ways, in addition to using an electricity-demanding MVHR which they had not previously used (although it is worth stressing that MVHRs tend to recover nine times more energy than they demand). Therefore, despite heating (and hot water, as they are measured together) fuel demand commonly being considerably lower than the average demand for existing or new buildings, electricity demand rarely changes significantly (Figure 6.3).

There have been numerous studies that have monitored the energy performance of Passivhaus buildings in recent years. See Table 6.3 for some UK examples, which provides insight into the first wave of Passivhaus projects in a particular country. A common finding of such, but not necessarily all, studies has been that the modelled (using PHPP software) energy demand of the building is usually lower than the actual monitored post-occupancy energy demand (see Chapter 4 for more on the energy performance gap). Nevertheless, it is commonly argued (e.g. McDonagh-Greaves, 2014) that the absolute size of this performance gap is much smaller for Passivhaus buildings, compared to other design and construction approaches. This is because its super insulation and airtightness reduces heating fuel demand to such an extent that any additional demand (in proportional terms) is said to be relatively negligible in absolute terms. For example, a building predicted to demand 10,000 kWh per year, but with a 10 per cent energy performance gap (actual demand 11,000 kWh) will be affected much more by the performance gap in absolute terms than a building predicted to demand 1,000 kWh per year with a similar 10 per cent performance gap (actual demand 1,100 kWh). To extend this argument further, this tends to mean that even those Passivhaus buildings with an energy performance gap and a larger energy demand still use much less energy than other (more conventional) building types. It is such evidence that is commonly used by Passivhaus advocates to argue that the standard will guarantee lower energy demand regardless of how occupants use the technologies and behave inside their particular buildings (e.g. Passivhaus Trust, 2011).

As discussed in Sections 6.2.2 and 6.2.3, Passivhaus buildings achieve lower heating fuel use through minimising building fabric and ventilation heat losses, by stipulating U-value and airtightness requirements that are much more stringent than the current building regulations of most countries. As such, Passivhaus buildings are able to successfully harness energy from solar gain and incidental gains for space heating, meaning that very little heating fuel is required to achieve 'comfortable' indoor temperatures (see Figure 6.1 for the modelling assumptions behind this). It is the significance of incidental gains, the passive accumulation of heat from everyday actions in the buildings (e.g. appliance usage, body warmth), that we argue is one of the most salient features of Passivhaus buildings. Indeed, low-energy buildings that make the most of solar gain have been constructed for many decades (Balcomb, 1992).

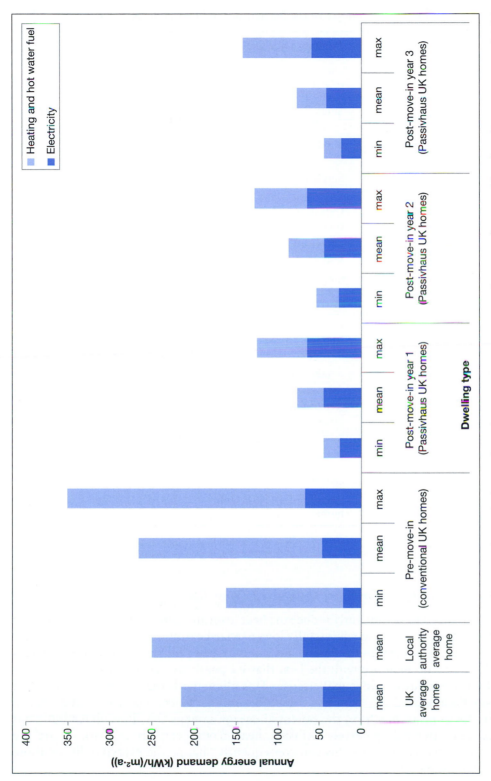

Figure 6.3 Comparison of annual energy demand of Passivhaus homes, tenants' previous homes and UK and local authority averages (kWh/ (m²·a)) (2011–2014).

Source: based on Ingham (2014); DECC (2014).

Table 6.3 Energy performance of a sample of UK Passivhaus (and EnerPHit) projects

Project name (location)	Building specification	Primary energy demand (kWh/(m²·a))		
		Modelled	*Actual*	*% diff.*
Camden Passivhaus *(London, England)*	Single-family house	99	119	+20
Crossway *(Staplehurst, Kent, England)*	Single-family house	55	57	+4
Denby Dale *(Yorkshire, England)*	Single-family house	87	110	+26
Green Base *(St Helens, Lancashire, England)*	Office and community building	107	118	+10
Lancaster Cohousing *(Lancashire, England)*	41 owner-occupied houses	81	77	−5
Plummerswood *(Innerleithen, Scottish Borders, Scotland)*	Single-family house	120	105	−13
Racecourse Estate *(Houghton-Le-Spring, County Durham, England)*	Four terraced bungalows for Gentoo Homes	103	231	+124
Stories Mews *(Camberwell, London, England)*	Single-family house	87	119	+37
Grove Cottage *(Hereford, England)*	Retrofit of single-family house	109	122	+12
Mayville Community Centre *(Islington, London, England)*	Retrofit of a nineteenth-century brick building to create a new community centre	120	70	−42
Totnes Passivhaus *(Totnes, Devon, England)*	Retrofit of a 1970s family home	68	76	+12

Source: Passivhaus Trust (2014).

6.4.2 *Improving heating performance through everyday life*

It is clear that we need to significantly reduce the heat loss rates (both fabric and ventilation) of our building stock if we are to successfully lower space heating energy demand. Whether or not one specifically adheres to the Passivhaus standard, the philosophy behind Passivhaus – that we should be doing more with the heat that we passively generate as part of taken-for-granted daily activities – does make sense. This aligns with the strong argument that technological design needs to better account for more efficient ways of using energy. However, by focusing so much on the potential energy savings that Passivhaus buildings can achieve (e.g. through passively utilising heat from everyday activities), are we giving adequate attention to how this can consequently change the everyday life of those occupying the buildings?

The Passivhaus requirement that internal temperatures should not exceed 25 °C for more than 10 per cent of the year is assessed in theoretical terms using the PHPP model, prior to actual construction and occupation. However, modelling everyday life is not a straightforward task due to its often unpredictable and dynamic nature (Higginson *et al.*, 2014). Thus, if we cannot accurately model everyday life, then there may well be problems in accurately modelling internal temperatures in Passivhaus (or similarly constructed) buildings. This may be especially true if people's everyday activities change as a direct consequence of those activities having a thermal impact. This has implications for how they manage those thermal impacts. Foulds (2013) discusses how occupants may reduce their level of multi-tasking, so as to avoid overheating. For example, perhaps avoiding having the cooker on while one household member vacuums and another washes and dries their hair (all appliances-related heat), prior to then hosting a large number of friends (body heat).

In relying on everyday life to help heat a building, the building's heating systems need to be designed differently in comparison to more conventionally designed buildings. It is wrong to say (as many do) that Passivhaus buildings have no heating systems, they merely have no need for a conventional heating system of distribution pipes and radiators, since the necessary heat supply can be delivered by the supply and extraction of cooled/heated air (see Chapter 7 for more details on MVHRs). A MVHR system is a good example of how new technologies can prove to be confusing for some occupants, who attempt to use it on the basis of knowledge acquired by previously using very different technologies. For example, some occupants have been found to be controlling the MVHR with the (wrong) assumption that warmth is achieved through turning controls 'up' (as you would do with a thermostat), not 'down' (Foulds, 2013). Yet, the ventilation rate would often need to be lowered to help warm the building up (as it allows for the passive build-up of occupant-generated warmth) and, assuming there is not a heat wave where outside is actually warmer than inside, ventilation may also need to be increased to help cool the building (as it removes the warm internal air more quickly).

6.4.3 Professional expertise and practical knowledge

Many colleagues, in industry and academia, commonly argue that Passivhaus buildings are simpler to use, as they focus on a good-quality build and put the 'fabric first', in comparison to buildings that employ lots of eco-gadgets and renewable technologies. Such perspectives, first, underplay the significance of how everyday life helps to heat these buildings (Section 6.4.1), and, second, suggest an over-emphasis on building occupants ('end-users'), when there may also be considerable struggles within industry (different sources of energy and technology 'users') in adapting to Passivhaus design and construction.

As implied by the previous section, many people may require time to allow them to adapt and become used to the new Passivhaus technologies that surround them. While this may be relatively short-term and usually only until the occupants become familiar with such technologies, it is important to remember the vital role that practitioners play in supporting occupants and easing them through this period of adjustment. Indeed, the 'handover' of a new building (particularly a low-energy building, such as one built to Passivhaus standard) is not merely something that occurs as the keys of the property are handed over to the new occupant on move-in day. Handover is a process of support, whereby relationships are built and maintained between those with and without knowledge of the (Passivhaus) technologies. Thus, we argue that handover should arguably begin before move-in day and should also finish perhaps a year after move-in day. We suggest that it may well be appropriate for support to be maintained until at least the end of the 'defects liability period' (i.e. the period of time that the contractor is contractually obliged to repair defects in their work). When handover procedures are adjusted, so they are appropriate for Passivhaus homes instead of 'normal' homes, we suggest that housing professionals take note of the recommendations detailed in Table 6.4.

Table 6.4 Recommendations for improving occupiers' handover

Recommendation	Details
Give guidance at appropriate times	New occupants should not be bombarded with information on move-in day because they will have other concerns (e.g. where to put the furniture). Thus, give some basic information about the low-energy technologies on move-in day, and return at a later date for a more detailed discussion.
Communicate with the other professionals involved	It is vital that all the relevant industry professionals communicate among themselves prior to issuing advice to the occupants. Otherwise, the occupants are likely to be confused and potentially mistrusting of any guidance being given. For instance, whose responsibility is it to maintain the MVHR and change its air filters?
Make explanations 'hands on'	The occupants are likely to significantly benefit if explanations about technologies are made more interactive. For example, a household could be taken on a technology tour of their new home within which they have the opportunity to play with MVHR or heating controls (as opposed to simply being given a manual or an expert talking to them about the subject).
Provide ongoing support	There is an adjustment period that occupants must go through as part of adapting their lives to new surroundings and establishing new routines. This includes learning how best to use their building (and its associated technologies) in the way that best suits their needs and wants. As such, support should ideally be maintained (or at the very least, available, perhaps in a reactive capacity) over a period of time, and should not be limited to only a few days around move-in day.
Avoid negative messages	To help maintain and harness engagement, many argue that positivity is important. For instance, telling occupants that it may take them up to two years to fully learn how to live in a Passivhaus building could contribute an early disengagement (e.g. 'is it worth even bothering?').
Be seasonally relevant	How a Passivhaus building is used depends on the seasons to a certain extent. For instance, in the summer, to avoid overheating windows may be opened overnight so as to 'purge' all the warmer air that has been accumulated through the day. Alternatively, in the winter, to take advantage of a Passivhaus building's relative airtightness, the windows may be closed to keep the passively accumulated heat in the building for as long as possible. Therefore, handover should be sensitive to such differences, as well as considering doing a short return visit at a change in the seasons, so as to reflect the change in occupants' needs.
Remember it is the household that consumes energy	The efforts of handover should not be focused on only one (potentially interested) household member. A building's technologies are used (and thus energy demanded) on the basis of negotiations and exchanges that occur between all household members. For instance, while it could just be one household member that understands how to control the heating system, how those controls are set will usually depend upon the thermal comfort preferences of the whole household.

(Continued)

Table 6.4 (Continued)

Recommendation	Details
Do not forget non-Passivhaus guidance	While it is essential that appropriate Passivhaus-related guidance and support is provided, this should not be to the detriment of non-Passivhaus (or indeed non-low-energy) guidance.
Ensure guidance is relevant to everyday life	Some argue that however much the information that is given to occupants is perfected, it does not necessarily matter because they will not be paying attention to that information anyway. Indeed, instruction manuals and written guidance are often never read and may be lost. While to a certain extent this is true (i.e. optimised information provision is not a magic-bullet solution), occupants may perhaps pay more attention to guidance if that guidance is made more relevant to everyday life. For instance, rather than talking about what the MVHR is and how it works, it would be preferable to talk about how the MVHR can practically be used. For example, turn it up when people are visiting; use the boost function when showering or cooking.

Source: based on material from Ingham (2013).

Various maintenance tasks, which are also not required for conventional buildings (be they existing or new builds), are required for Passivhaus buildings. Perhaps the most important of these is changing the filters on the MVHR, without which the high efficiency of the unit will be lost and the building's internal air quality may also be detrimentally affected. It is recommended that filters are changed 2–4 times per year (Price and Brown, 2012), depending on whether you live in a rural or urban location, respectively. This said, depending on the exact rural location (e.g. near to a farmer's field with its agricultural pollutants), then rural buildings may need new filters just as frequently as urban buildings. It is thus advised that each building's filter maintenance regime is assessed on a case-by-case basis, which would include the physical examination of used filters. See Section 7.3.2 for more details of mechanical ventilation.

Evidence currently suggests that when problems do arise (mainly due to technological breakdowns), many building professionals struggle to know exactly what to do. For example, Foulds (2013) provides examples of occupants suffering with technological breakdowns, only for maintenance contractors to come out and then not know what to do, including after telephoning colleagues and even asking the occupant themselves for advice. Maintenance staff often had to return for multiple visits before a problem or breakdown could be rectified, because they were simply not familiar enough with a Passivhaus building's technologies. However, it is important to note that such technological issues were not driven directly by the Passivhaus standard itself, but instead were simply symptomatic of the building industry's inability to successfully integrate new low-energy technologies (that just happened to be associated with this particular Passivhaus project) into the built environment.

Construction teams delivering Passivhaus buildings may also have to change the way they go about their work. For instance, Macrorie *et al.* (2015) reveal that despite significant supply chain improvements in recent years, they have not improved to the extent whereby Passivhaus components and products can always be easily and competitively sourced, quite often having to be imported from more established markets, such as Austria or Germany. Macrorie *et al.* (2015) also discuss in their Passivhaus case study that construction professionals struggle with the attention to detail that is required, particularly when trying to achieve the Passivhaus standard's stringent airtightness requirements. For example, while traditional bricklaying can be a little haphazard (providing that the external wall is flush on its exterior face), a Passivhaus external wall needs to be more carefully constructed (so that the wall is flush on both interior and exterior faces) for airtightness purposes. Furthermore, a reoccurring theme in Schoenefeldt's (2014) e-book of UK Passivhaus case studies is the skills gap in the UK industry, meaning that professionals often struggle to tackle the quality-related challenges that Passivhaus presents. It is therefore unsurprising that conversations with colleagues in industry show there is often on-site tensions and confrontations regarding how to practically achieve airtightness (and also if airtightness is even necessary).

Furthermore, from conversations with colleagues, we have also heard stories of client briefs stipulating additional gas pipework and/or electrical wiring in parts of the building where it is not actually required (at an additional cost too) because of a fear that the Passivhaus design concept may not provide occupants with sufficient warmth. Thus, such pipework and wiring essentially acts as a contingency strategy. The key point here is that many built environment professionals, who currently lack expertise in this area, are fearful as to whether significantly different design and construction approaches (such as Passivhaus) can deliver. It is thus perhaps of no surprise that relevant professional bodies, such as the Passivhaus Trust, are actively trying to issue guidance on how to design, construct, maintain and support people occupying Passivhaus buildings (Taylor and Raffiuddin, 2013; McLeod, *et al.*, 2012a; 2012b).

As this section implicitly emphasises, the Passivhaus research literature, and indeed the energy demand literature more widely, often over-focuses on the 'end-user' (the occupant operating the technologies), but there are numerous other types of 'users' that are using the technologies (e.g. in their design, construction and maintenance). These other 'users' arguably need more attention, especially when research indicates that professionals need to adapt to future built environment technologies, just as the occupants are. Such issues are not exclusive to the Passivhaus standard, but are relevant to the manufacture, installation, certification, management, maintenance and operation of all newly emerging technologies. Moreover, such struggles are arguably inevitable as professionals learn to adapt their working practices to include new technologies: they cannot have experience and a working knowledge of a new built environment technology until they have actually experienced it.

6.4.4 *Reflecting upon the Passivhaus standard as an example of an energy-efficiency solution*

The Passivhaus standard is just one example of numerous building codes and standards that are currently proposed or already in use; the majority of these tend to focus on improving building energy efficiency. Indeed, many praise Passivhaus for prioritising energy efficiency technologies, in particular super insulation and airtight construction, over renewable energy technologies, which are often referred to as unnecessary 'eco-bling' (King *et al.*, 2012; Burrell, 2014; Tofield, 2012). As King (in Jha, 2010) puts it:

> if you build something that is just as energy-hungry as every other building and then put a few wind turbines and solar cells on the outside that addresses a few per cent of that building's energy consumption, you've not achieved anything.

It is common sense to invest in energy efficiency before renewable energy technologies, because reducing energy demand results in reducing the size of the system required and hence the total cost, as less energy needs to be generated (see Chapter 2 for details of the energy hierarchy). However, one could argue the exact same point for energy conservation needing to be prioritised over energy efficiency projects, because energy efficiency (in exactly the same way as renewables do) still fundamentally allows building occupants to continue to demand energy unsustainably (e.g. leaving lights on unnecessarily). Efficiency improvements essentially give occupants a 'free pass' to maintain the status quo. Moreover, if anything, energy efficiency projects can actually incentivise more energy-demanding patterns of use, as improved efficiency makes it cheaper to do so (see Section 4.8.3 for details of the rebound effect). For instance, it is very likely that Passivhaus occupants will experience higher indoor temperatures than they were previously used to. This is most likely why the PHI states that Passivhaus is not only an energy-efficiency standard, but also a (thermal) comfort standard.

It is clear that technological change is intertwined with social change (Guy, 2006). Therefore, by focusing so much on heating-related energy efficiency, it is perhaps unsurprising that such technologies have the potential to re-define social conventions around thermal comfort, in such a way that they 'ratchet up' our energy demand (i.e. as technology develops, a new standard of living, and pattern/level of energy demand, becomes regarded as the new minimum). As Shove (2003, p. 194) remarks, in the very similar context of redesigning homes for air conditioning, 'by building this expectation (here, with regard to thermal comfort) into the fabric of the property itself, consumption of energy was inevitably ratcheted up'. Indeed, minimising heat loss by creating an internal–external barrier that essentially standardises the internal environment year-round (meaning that seasons, for instance, have little impact internally), Passivhaus buildings can change how occupants perceive and aim to achieve thermal comfort (e.g. normal not to wear a jumper indoors during the winter). This is markedly different to other proposed approaches to managing thermal comfort. For example, adaptive comfort proposes that occupants need to be more actively involved in managing their thermal comfort and that we need variation (not standardisation) in indoor conditions (Nicol and Stevenson, 2013). It is vital to consider such issues as the design intentions behind our current and planned buildings leave a lasting legacy for our everyday life (and thus our energy demand patterns) in the future.

Therefore, there is an argument that we need to focus more directly on tackling our very need for energy (i.e. energy conservation), rather than ignoring the root causes of demand. It could be argued that Passivhaus jumps straight to energy efficiency (Stage 2 of the energy hierarchy), and thus bypasses energy conservation (Stage 1) (see Section 2.3 for more about energy conservation). Furthermore, since it is difficult to see how much a building standard could feasibly get to the heart of energy conservation, there is an indication that we need to consider other solutions too.

6.5 Case studies

6.5.1 Case study 1: Brussels-Capital region, Belgium (EnEffect, 2012)

Rationale

This case study provides insights into how the Passivhaus standard's requirements can be incorporated into the construction standards for a whole region. There have been numerous calls internationally, at both local and national levels, to make voluntary standards such as Passivhaus mandatory, and hence this is an interesting example of success which needs to be reflected upon.

Project background

- Location: Brussels-Capital region, Belgium
- Building type: changes to construction standards are applicable to both residential and commercial buildings
- Tenure: applicable to all tenures
- Year of construction: legislation is applicable from the start of 2015 onwards.

This case study is sourced from the EU-funded PassReg project, the principal objective of which is to drive energy-efficient construction throughout the EU. In particular, the promotion of the Passivhaus standard is a core activity, which the project primarily addresses by focusing on sharing best practice internationally. Emphasis is placed on both 'front runner' regions and those regions aspiring to become front runners in the future. As part of this work, the project compares the successes of Brussels to the context-specific circumstances of other regions – for example, Hanover has taken decades, rather than a few years (like Brussels), to reach the 'front runner' status (for more details on other case studies, see EnEffect (2013)).

Approach

The project collected the information through discussions with the appropriate ministers and officers in Brussels, as well as through a detailed review of relevant past documents (e.g. Brussels political meetings, policy documents, press releases, presentations).

Findings

In 2014 there are 860 new Passivhaus projects under construction in Brussels, which will provide over 2,300 Passivhaus homes to Brussels' social housing stock. Yet in 2007, the Brussels-Capital region did not have one single Passivhaus building due to a perception by local industries and architects, as well as the wider Brussels public, that low-energy buildings were merely a luxury item. Therefore there has been a considerable change in recent years.

Such has been the momentum of the Passivhaus standard in Brussels, and indeed low-energy construction more generally, that it has culminated in the region committing to a method of construction that is very similar to the Passivhaus standard. Indeed, it has been legislated (Brussels Hoofdstedelijk Gewest, Ministerie Van Het Brussels Hoofdstedelijk Gewest N. 2011-2445) that from the beginning of 2015 onwards, all new construction (be it housing, offices or schools, for instance) must meet a mandatory 'passive' standard that is only slightly different to the Passivhaus standard. Specifically, the legislation requires:

- a net cooling/heating requirement of less than 15 kWh/(m²·a) (cooling requirement only relevant for offices and schools);
- an airtightness requirement of 0.6 ach;
- an overheating requirement of 25 °C for no more than 5 per cent of the year (or 5 per cent of working time for offices and schools);

- a primary energy demand limited to:

  o   45 kWh/(m^2·a) for housing (heating, hot water, ventilation, pumps and fans);
  o   90 – 2.5 × surface area to volume ratio (A/V) kWh/(m^2·a) for offices and schools.

Although the PHPP software will be central, it is not the Passivhaus Institute in Germany that will be certifying the buildings themselves. Instead, it will be the Belgian-based sister non-profit organisations that will be certifying the buildings against the legislative requirements. It is clear, therefore, that the Brussels-Capital region is boldly leading the way, as a frontrunner in Passivhaus-related legislation.

A key driver behind the adoption of the standard is political action, particularly at the regional level. While both the federal and regional governments technically share responsibilities for environmental issues, in reality it is the three Belgian regional governments that tend to take ownership of environmental governance (and, as such, low-energy construction policies too). Thus, it was a significant moment when the Brussels-Capital region government not only became convinced about the merit of low-energy construction, but actually made a clear political commitment to it. In making this public political move, the regional government were aided by useful discussions and exchanges of best practice with other EU regions (in particular Franche-Comté, France) that were also attempting to spearhead stringent low-energy legislation.

The Brussels regional government did, especially initially, face some concern and opposition from the public and industry with regard to the low-energy construction requirements that they were planning to make mandatory. In successfully addressing such concerns, the authorities focused on using various support measures, rather than introducing more energy policies as part of a harder-line approach. For instance, householders living in Passivhaus homes became ambassadors of the Passivhaus standard. These householders helped to demonstrate that support (e.g. regarding how to use Passivhaus technologies) did indeed exist, as well as emphasising that Passivhaus buildings were not simply a luxury construction feature reserved for the rich, as most of Brussels' Passivhaus households are from low-income backgrounds. To complement these efforts, performance monitoring results of past Passivhaus building projects were publicised, so as to demonstrate past successes and that this was a credible and cost-effective direction for the region to take. In particular, the Brussels' Calls for Exemplary Buildings (in 2007, 2008 and 2009) – which is an annual call where the regional government supports exemplary building projects through financial, technical and marketing assistance – acted as a test bed for the Passivhaus standard, and confirmed the Passivhaus standard 'is fully accessible and does not lead to major increased costs . . . in new construction and sometimes even renovation' (Deprez, 2011, p. 5).

These legislative actions have positioned the Brussels-Capital region as a reference point for other regions in the world that also aspire to Passivhaus-related legislation. For example, Binns *et al.* (2013) reflect on the experiences of Brussels in drawing out recommendations for achieving something similar in the East of England region.

Key messages

- Positive experiences (e.g. achieving high energy performance at low hassle and affordable cost) of how successful Passivhaus developments convinced stakeholders of the merits of Passivhaus construction.

- The initial fear and resentment of the standard was tackled directly through stakeholder engagement.
- A welcoming socio-political landscape was pivotal in the standard being suggested in legislative terms, let alone actually accepted.

6.5.2 Case study 2: constructing to Passivhaus in contrasting climatic zones (International Passive House Association, 2014c)

Rationale

This study illustrates the potential for designing and constructing Passivhaus buildings in a range of different climates, beyond that of Western or Central Europe, which is where Passivhaus originated and has developed the most to date. The Passivhaus standard is increasingly being sold as an energy-efficiency solution of global relevance, and hence this case study can help to explore if indeed this is true.

Project background

- Locations:

    o *Cold climate:* Yekaterinburg (Russia)
    o *Sub-tropical climate:* Tokyo (Japan), Shanghai (China)
    o *Hot and dry climate:* Las Vegas (USA)
    o *Hot and humid climate:* Dubai (UAE)
    o *Tropical climate:* Salvador (Brazil), Mumbai (India)

- Building type: residential
- Tenure: irrelevant, as the study is about the feasibility of construction
- Year of construction: were not actually constructed, but modelling was based on technologies available in 2012–2013
- Size: one two-storey reference home with an internal floor area of 120 m²
- Construction: each reference home had thermal properties that would enable Passivhaus certification.

The aim of the project was to systematically and scientifically assess the principles and true potential of practically constructing Passivhaus buildings in other climatic zones. Specifically, climatic zones that contrast with Central and Western Europe (e.g. Germany), where most Passivhaus buildings have been constructed to date.

   The work presented here actually forms part of two separate studies: 'Passive House for different climate zones' (Schnieders *et al.*, 2012) and 'Passive Houses in tropical climates' (Schnieders and Grove-Smith, 2013). In particular, we draw on the overview of these two studies from the Passipedia website (International Passive House Association, 2014c).

Approach

The modelling exercise compares an identical reference home (two-storey end-of-terrace, with 120 m² floor area), constructed to the thermal, airtightness and energy requirements of the Passivhaus standard, but located in eight different climatic zones. The various climatic zones were chosen so as to provide comparisons between dry and humid, as well as hot and cold, climates, as there are concerns about how best to deal with moisture-rich air and the extent of heating/cooling required.

Findings

The main finding indicates it is technically feasible to construct Passivhaus buildings in all eight locations, regardless of their specific climatic conditions.

Interestingly, there are no significant problems in Passivhaus being applied in the hot climatic contexts, defined as having outdoor temperatures of up to 40°C, with the Passivhaus components that are already on the market managing to suffice. However, there are problems constructing Passivhaus buildings in very cold climates, which were defined as having daily average temperatures of as low as −70°C, with negligible solar insolation (although Yekaterinburg temperatures are not this cold, the modelling aimed to explore climatic extremes). In particular, it is the currently available Passivhaus-certified windows that are deemed not thermally adequate for these extreme climatic conditions, and thus any prospective Passivhaus buildings would require customised components. As such, achieving the considerably lower heat loss rates required by the Passivhaus standard is likely to be uneconomic (but nevertheless technically achievable).

Constructing Passivhaus buildings in the tropics (here, Salvador and Mumbai were considered) is regarded as a unique case due to very minimal temperature changes occurring across the seasons. The consequence of this is that while no heating is required, the energy demand of cooling could be significant. A suggested way to reduce the cooling demand is to focus on internal thermal insulation, as opposed to external, which allows for easier installation of exterior shading systems and deeper window jambs, thereby reducing the amount of solar gain. Internal insulation will also have a beneficial effect on the moisture balance of the structure, which needs to be tackled to ensure longevity of a building in the tropics. An external vapour retarder and the active dehumidification of indoor air are also recommended to protect the building structure against external moisture risks.

Other notable issues that differ between achieving the Passivhaus standard in Central Europe, as opposed to in other climatic zones, include dealing with: the risk of frost in building foundations and in the MVHR; a potential need for separate cooling and dehumidification; and condensation on exterior surfaces in tropical climates (for more details, see Schnieders *et al.* (2012) and Schnieders and Grove-Smith (2013)).

The reports also made it clear that one design approach, which may be perfect for one particular country, may be completely inappropriate for another country with a different climate. In addition, they noted that exactly how the Passivhaus concept is operationalised in one particular country not only depends on the climate, but also upon the local building traditions and preferred aesthetics. This is, of course, something that the Passivhaus standard implicitly encourages, as it does not dictate a particular design or construction approach, instead only stipulating what those chosen approaches must achieve in terms of, for instance, energy demand and airtightness.

Key messages

- It is technically possible to construct buildings to the Passivhaus standard in a range of different climatic zones.

- The customising of Passivhaus components (e.g. windows) in very cold climates may, however, make it uneconomical.
- Internal insulation is recommended for tropical climates, so as to help manage external moisture risks and also allow for the installation of shading systems on the exterior.
- There is no one way to design and construct a Passivhaus building, as it depends not only on climate, but also on local conventions and preferred aesthetics.

6.5.3 Case study 3: Ditchingham Passivhaus-certified housing development, UK (Richards, 2013)

Rationale

As is a common theme throughout this book, it is important to remember that the 'success' of any technology depends on how exactly that technology is taken up and used by its occupants. Therefore, with a particular focus on the experiences of the new occupants, this case study explores the challenges of 'handover' (from contractor to occupants).

Project background

- Location: Ditchingham, Norfolk, UK.
- Building type: residential
- Tenure: social renters and shared owners
- Year of construction: 2011–2012
- Size: 14 homes
- Construction approach: brick/block cavity construction.

This 14-dwelling affordable housing development is the second Passivhaus development constructed for Hastoe Housing Association (Passivhaus Trust, 2012c). The first development, built at Tye Green Wimbish, is the basis for a case study in Chapter 4. This second development, which consists of a mix of one-, two- and three-bed flats, bungalows and houses, has a combination of shared ownership and social renting tenures. The architect is Parsons + Whittley and the contractor (construction team) is Keepmoat. With the construction cost totalling £1,454/m^2, calculations from Hastoe's proceeding Passivhaus development suggested construction costs are 12 per cent higher than CfSH Level 4 housing (Ingham, 2013), although it is very likely that Ditchingham will be slightly less costly due to it being the second time the project team had worked on a Passivhaus new build. More details on the case study can be found at Passivhaus Trust (2013b).

Approach

Household interviews were conducted with residents soon after move-in. The structured interviews mainly consisted of the researcher using a pre-prepared questionnaire, which meant that (other than a small space for additional comments) there was little opportunity to explore households' experiences that may not have been previously anticipated. In addition, building monitoring was undertaken as part of investigating the energy performance of the homes.

Findings

Monitoring data indicated that all the homes had a higher electricity demand than anticipated, with half being more than double that modelled in PHPP. This is an interesting point as it supports the earlier argument regarding how the Passivhaus standard may struggle to tackle electricity demand (Section 6.4.1), in addition to it reinforcing the discussion in case study 3 of Chapter 4 (Section 4.10.3) regarding modelling assumptions not equalling reality. Nevertheless, here we focus on the handover procedures and support offered to households occupying these new build Passivhaus homes – which would have inevitably contributed to these energy demand trends.

Through the interviews, the households report that the housing association provided a 'handover meeting', which was conducted at move-in and was designed mainly with the purposes of educating the occupants on how best to use their new home's technologies (in particular, the heating and ventilation systems). However, only 30 per cent of the households said that they actually attended the handover meeting. This was said to be mainly due to the meeting being inconveniently timed (e.g. clashing with work). As such, many households had no handover advice, with one shared ownership household even receiving their handover advice from an estate agent, who knew next to nothing about Passivhaus homes.

However, discussions with the housing association contradict what the households were reporting. The association asserts that every single household had a pre-handover meeting, as well as an additional handover meeting on move-in day. Furthermore, the households are said to have had two other opportunities to ask questions and receive guidance soon after move-in, which is not offered to their 'normal' housing development tenants, but apparently no households took advantage of this. The fact that there is a differing version of events between the housing association and the households suggests that their expectations of handover were not aligned, and that communication (particularly regarding each party's wants and needs) could be improved. However, perhaps the key point here is that, whatever the underlying issues, handover procedures were not sufficient in providing the households with the skills to operate Passivhaus technologies.

But even if improvements to the handover process were adopted and every household did participate in the various handover meetings, it does not necessarily mean that it will either improve their technological understanding and/or change how they actually use their new Passivhaus home. For example, only two households ever consulted the handover packs they were given at move-in. It should not be assumed that improving how the handover packs are written will guarantee energy saving, because improving the households' knowledge will not necessarily ensure 'correct' (low energy) operation of their home.

This whole situation would have contributed to most of the households not understanding how their everyday actions (e.g. appliance usage) relate to the internal temperature of their home. The way this handover has been organised is initially surprising, given that this was the housing association's second Passivhaus development, and so could have been expected to utilise experience from the handover at their first development. However, it is perhaps more understandable given that there were different project officers involved, due to the internal decision to 'share around the Passivhaus responsibility' as part of helping to up-skill everyone in their workforce.

Key messages

- Electricity demand was significantly higher than that expected through the PHPP model.
- Households lacked an understanding of how to use the Passivhaus technologies and, in particular, how their actions (e.g. appliance usage) could change internal temperatures.
- The residents and housing association disagree over how exactly handover had been organised, suggesting they had different expectations and that communication could have been improved.
- While improvements to handover are vital, it should not be regarded as a magic-bullet solution. An optimised version of handover will not guarantee the residents will learn anything, let alone choose to act on that knowledge and save energy.

6.6 Summary

Passivhaus, which originated in Germany in the early 1990s, is an energy-efficiency standard for all types of buildings. Since that time, there has been considerable growth internationally and the standard has gathered increasing support in industry and policy circles. Passivhaus buildings tend to be super-insulated and airtight, with an MVHR unit installed to ensure good air quality is maintained, as this cannot be achieved through natural ventilation in cold weather.

The standard requires that energy demand for space heating/cooling does not exceed 15 kWh/(m²·a) (or, alternatively, the heating load must be no higher than 10 W/m²). In addition, the total primary energy demand of 120 kWh/(m²·a) must not be exceeded. But since certification is based on the modelled outputs of the PHPP tool, then inevitably actual energy demand is likely to differ somewhat. Nevertheless, energy performance monitoring studies indicate that significant energy savings are usually achieved and the total energy demand of 120 kWh/(m²·a) is mostly realised. When this has not been achieved, energy demand is still considerably lower than that of a conventional home. However, while significant heating fuel savings are evident, Passivhaus occupants rarely seem to reduce their electricity demand compared to their previous (non-Passivhaus) home's energy bills.

Many of the main challenges associated with adapting to the Passivhaus-related technologies are often just symptomatic of moving technologies from the niche (where it is emerging and unknown) to the mainstream (where it is established and familiar). For example, for occupants, most of the practical obstacles relate to a lack of experience, such as with the MVHR and the fact that in a Passivhaus building everyday actions have an impact on thermal comfort. It is important to note, though, that here we are referring to the inexperience of both the building occupants and the relevant housing professionals (e.g. landlords, construction teams, architects, project managers), as they too may have to adjust how they go about their working day, if they are to ensure that Passivhaus certification is achieved and/or the occupants are adequately supported.

More broadly, the Passivhaus standard is an example of an (energy-efficient) technological fix, in that the assumption behind its supporting rhetoric is that 'getting the technological design right' will almost guarantee energy savings. While the energy performance chapter (Chapter 4) indicates that there are no such guarantees, this chapter also implicitly emphasises that if energy demand is only a consequence of the way we live our lives, then changing the energy-demanding technologies that surround us will inevitably impact upon how we live (which in turn may also influence energy demand). The Passivhaus standard is an interesting illustration of how a building standard could be standardising, potentially for the better – but possibly for the worse, depending on your viewpoint – our internal environment and many aspects of our everyday lives (cf. Shove and Moezzi, 2002). Indeed, if adhered to more widely (with, for example, more regional governments making it a mandatory requirement), then the Passivhaus standard is likely to change thermal comfort social conventions (e.g. what it means for you to be 'sufficiently' warm and comfortable) and thereby the reasons for why we are consuming energy for space heating. Nevertheless, it is important to note that comfort conventions have been significantly evolving over recent decades and are likely to continue to do so, whether or not more Passivhaus buildings are constructed. Therefore, while it is clear that the Passivhaus standard has the potential to considerably reduce building energy demand, we need to (as with any low-energy or carbon technology) consider how those technologies may influence and be influenced by occupant activities.

References

Anderson, R. (2013) Homes: warm welcome. *Guardian*. www.guardian.co.uk/lifeandstyle/2013/feb/08/homes-warm-welcome-passivhaus, accessed 9 February 2013.

Atanasiu, B., Boermans, T., Hermelink, A., Schimschar, S., Grözinger, J., Offermann, M., Thomsen, K.E., Rose, J. and Aggerholm, S.O. (2011) *Principles for Nearly Zero-energy Buildings: Paving the Way for Effective Implementations of Policy Requirements*, Ecofys, Brussels. www.ecofys.com/files/files/ecofys_bpie_2011_nearlyzeroenergybuildings.pdf, accessed 5 June 2015.

Badescu, V. and Sicre, B. (2003a) Renewable energy for passive house heating: II. Model. *Energy and Buildings*, 35(11), 1085–1096.

Badescu, V. and Sicre, B. (2003b) Renewable energy for passive house heating: Part I. Building description. *Energy and Buildings*, 35(11), 1077–1084.

Balcomb, J.D. (1992) Introduction. In: J.D. Balcomb (ed.), *Passive Solar Buildings*, MIT Press, Cambridge, MA.

Binns, B., Standen, M., Tilford, A. and Mead, K. (2013) *Be.Passive: Lessons Learnt from the Belgian Passivhaus Experience*. Centre for the Built Environment, Adapt Low Carbon Group, Norwich. www.passivhaustrust.org.uk/UserFiles/File/BePassive Report.pdf, accessed 5 June 2015.

Boardman, B. (2012) *Achieving Zero: Delivering Future-Friendly Buildings*. Environmental Change Institute, University of Oxford, Oxford. www.eci.ox.ac.uk/research/energy/achievingzero, accessed 5 June 2015.

Bootland, J. (2014) UK Passivhaus uptake: The role of the Passivhaus Trust. *UK Passivhaus Conference 2014*, Passivhaus Trust, 16–17 October.

Brussels-Capital region (2011) Brussels Hoofdstedelijk Gewest, Ministerie Van Het Brussels Hoof dstedelijk Gewest N. 2011-2445. www.emis.vito.be, accessed 12 December 2014.

Burrell, E. (2014) Zero-carbon buildings? It's the wrong target. *Architecture in the Anthropocene*. http://elrondburrell.com/blog/zero-carbon-buildings-wrong-target/?utm_content=bufferf21c0&utm_medium=social&utm_source=twitter.com&utm_campaign=buffer, accessed 27 October 2014.

Cox, P. (2005) Passivhaus. *Building for a Future*, 15(3), 16–20.

Cutland, N. (2012) *Lessons from Germany's Passivhaus Experience*. IHS BRE Press, Milton Keynes.

DECC (2014) *Energy Consumption in the UK*, Department of Energy and Climate Change, London.

DECC (2010) Chris Huhne's speech to the Passivhaus conference. www.decc.gov.uk/en/content/cms/news/ch_passivhaus/ch_passivhaus.aspx, accessed 11 October 2010.

Deprez, B. (2011) *Brussels: From Eco-building to Sustainable City*. Department for the Environment, Energy, Urban Renewal and Welfare, Brussels-Capital Region, Brussels. http://documentation. bruxellesenvironnement.be/documents/BxlVilleDurable_ANGL.PDF, accessed 5 April 2015.

Earley, K. (2013) Hastoe Group lays the blueprint for sustainable communities. *Guardian*. www.guardian. co.uk/sustainable-business/hastoe-group-blueprint-sustainable-communities, accessed 17 May 2013.

EnEffect (2013) *Passive House Regions: A Guide to Success*. Passive House Regions with Renewable Energies (PassREg). www.passreg.eu/index.php?page_id=334, accessed 5 June 2015.

EnEffect (2012) The success model of Brussels: Case study. In: *Passive House Regions with Renewable Energies – Task 2.1.2: Describe the Critical Factors of Existing Success Models in Front Runner Regions*, IEE PassREg, Brussels. www.passreg.eu/index.php?page_id=334, accessed 5 June 2015.

Feist, W. (2013) *EnerPHit and EnerPHit+i: Certification Criteria for Energy Retrofits with Passive House Components*, Passive House Institute, Darmstadt. www.passiv.de/downloads/03_certification_ criteria_enerphit_en.pdf, accessed 5 April 2015.

Feist, W., Schnieders, J., Dorer, V. and Haas, A. (2005) Re-inventing air heating: Convenient and comfortable within the frame of the Passive House concept. *Energy and Buildings*, 37(11), 1186–1203.

Ford, P. and Hill, I. (2011) *Energy Efficiency and Microgeneration in the Built Environment, Skills Research for Scotland*. Scottish Government, Edinburgh. www.gov.scot/Publications/2011/10/ 04142122/0, accessed 4 October 2011.

Foulds, C. (2013) Practices and technological change: The unintended consequences of low energy dwelling design. PhD thesis, University of East Anglia, Norwich, UK.

GBPN (2013) Finland. www.gbpn.org/databases-tools/bc-detail-pages/finland#Summary, accessed 17 June 2013.

Guy, S. (2006) Designing urban knowledge: Competing perspectives on energy and buildings. *Environment and Planning* C: *Government and Policy*, 24(5), 645–659.

Hawkes Architecture (2014) Grand Designs: The Arched Eco House. www.hawkesarchitecture.co.uk/ grand-design, accessed 12 December 2014.

Higginson, S., Mckenna, E. and Thomson, M. (2014) Can practice make perfect (models)? Incorporating social practice theory into quantitative energy demand models. *Behave 2014 – Paradigm Shift: From Energy Efficiency to Energy Reduction through Social Change, 3rd Behave Energy Conference*, 3–4 September, Oxford.

Ingham, M. (2014) *Wimbish Passivhaus Development: Performance Evaluation Executive Summary – July 2014*. Norwich: Linktreat. www.hastoe.com/page/760/Wimbish-passivhaus-performs— Hastoe-releases-results-of-two-year-study.aspx, accessed 5 June 2015.

Ingham, M. (2013) *Wimbish Passivhaus: Building Performance Evaluation Second Interim Report – March 2013*. Norwich: Linktreat. www.wimbishpassivhaus.com/WimbishInterimReport.pdf, accessed 5 June 2015.

International Passive House Association (2014a) Examples. Passipedia. http://passipedia.org/examples, accessed 18 September 2014.

International Passive House Association (2014b) Passive House legislation. www.passivehouse-international.org/index.php?page_id=176, accessed 17 December 2014.

International Passive House Association (2014c) The studies 'Passive Houses for different climate zones'/'Passive Houses in tropical climates'. Passipedia. www.passipedia.org/passipedia_en/basics/ passive_houses_in_different_climates/study_passive_houses_for_different_climate_zones, accessed 18 September 2014.

International Passive House Association (2013) The world's first Passive House, Darmstadt-Kranichstein, Germany. Passipedia. http://passipedia.passiv.de/passipedia_en/examples/residential_ buildings/single_-_family_houses/central_europe/the_world_s_first_passive_house_darmstadt-kranichstein_germany, accessed 18 September 2014.

Jenkinson, E. (2010) Is this the housing standard of the future? *The Independent*. www.independent. co.uk/property/house-and-home/is-this-the-housing-standard-of-the-future-2112893.html, accessed 21 October 2010.

Jha, A. (2010) Eco-bling and retrofitting won't meet emissions targets, warn engineers. *Guardian*. www.theguardian.com/environment/2010/jan/20/eco-bling-retrofitting-carbon-emissions, accessed 20 January 2010.

King, D., McCombie, P. and Arnold, S. (2012) *The Case for Centres of Excellence in Sustainable Building Design*, The Royal Academy of Engineering, London.

Krestovnikoff, M. and Poyntz-Roberts, M. (2013) Costing the earth: The house that heats itself. BBC Radio 4. www.bbc.co.uk/programmes/b01r5ln3, accessed 13 March 2013.

Macrorie, R., Foulds, C. and Hargreaves, T. (2015) Governing and governed by practices: Exploring governance interventions in low-carbon housing policy and practice. In: *Social Practices, Interventions and Sustainability: Beyond Behaviour Change*, Routledge, Abingdon.

McCabe, J. (2012) Airtight promise. Inside Housing. www.insidehousing.co.uk/eco/airtight-promise/6521847.article, accessed 18 May 2012.

McDonagh-Greaves, L. (2014) POE: The Performance Gap. Warm Associates. www.peterwarm.co.uk/poe-the-performance-gap, accessed 13 August 2014.

McGhie, C. (2008) The green light for your eco-ambitions. *Sunday Telegraph: Home & Living*. www.telegraph.co.uk/finance/property/green/3360532/Eco-homes-Green-light-for-your-eco-ambitions.html, accessed 23 February 2008.

McLeod, R., Mead, K. and Standen, M. (2012a) *Passivhaus Primer: Designer's Guide – A Guide for the Design Team and Local Authorities*, The BRE Trust, Watford.

McLeod, R., Tilford, A. and Mead, K. (2012b) *Passivhaus Primer: Contractor's Guide – So You've Been Asked to Build a Passivhaus?* The BRE Trust, Watford.

Mekjian, S. (2011) Passive House: Going global. International Passive House Association. www.passivhaustrust.org.uk/UserFiles/File/PHT 1st anniversary presentations/2011-06-17 PH Internationally SM.pdf, accessed 17 December 2014.

Nicol, F. and Stevenson, F. (2013) Adaptive comfort in an unpredictable world. *Building Research & Information*, 41(3), 255–258.

Official Journal of the European Union (2008) European Parliament resolution of 31 January 2008 on an Action Plan for Energy Efficiency: Realising the Potential (2007/2106(INI)). European Commission.

Passive House Institute (2014a) New update available for Passive House Planning Package PHPP, Passive House Institute, Darmstadt.

Passive House Institute (2014b) Press release: Designers and architects gear up for the future of construction. www.passivehouse-international.org/download.php?cms=1&file=2014_06_10_Passive HouseDesigner_PressRelease.pdf, accessed 10 June 2014.

Passive House Institute (2014c) The awards recipients and finalists. www.passiv.de/archpreis, accessed 11 November 2014.

Passive House Institute (2013) *Passivhaus Planning Package (PHPP) Version 8 Manual*, Passive House Institute, Darmstadt.

Passivhaus Trust (2014) Projects. www.passivhaustrust.org.uk/projects, accessed 14 December 2014.

Passivhaus Trust (2013a) Canolfan Hyddgen, JPW Construction. UK Passivhaus Awards 2013: non-domestic category. www.passivhaustrust.org.uk/UserFiles/File/UK%20PH%20Awards/2013/Posters/UKPHAwardsPoster_non-domestic_Canolfan%20Hyddgen.pdf, London, accessed 18 August 2015.

Passivhaus Trust (2013b) Ditchingham Passivhaus, Parsons + Whittley Architects. UK Passivhaus Awards 2013: social housing category. www.passivhaustrust.org.uk/UserFiles/File/UK%20PH%20Awards/2013/Posters/UKPHAwardsPoster_social%20housing_Ditchingham.pdf, London, accessed 18 August 2015.

Passivhaus Trust (2013c) *Passivhaus: An Introduction*. Passivhaus Trust, London. www.passivhaustrust.org.uk/UserFiles/File/PH%20Intro%20Guide%20update%202013.pdf, accessed 16 September 2014.

Passivhaus Trust (2013d) Passivhaus schemes: Up-to-date map of all current certified Passivhaus projects in the UK. www.passivhaustrust.org.uk/projects/passivhaus_projects_map, accessed 16 September 2014.

Passivhaus Trust (2013e) Underhill House. www.passivhaustrust.org.uk/projects/detail/?cId=11#.Ub2bqee86DR, accessed 16 June 2013.

Passivhaus Trust (2012a) Denby Dale Passivhaus, Green Building Store. UK Passivhaus Awards 2012: residential category. www.passivhaustrust.org.uk/UserFiles/File/Projects/Awards2012/Denby%20 Dale_Green%20Building%20Store.pdf, London, accessed 18 August 2015.

Passivhaus Trust (2012b) Hadlow College RRC, Eurobuild. UK Passivhaus Awards 2012: non-domestic category. www.passivhaustrust.org.uk/UserFiles/File/Projects/Awards2012/PHTAwardsPoster_Hadlow. pdf, London, accessed 18 August 2015.

Passivhaus Trust (2012c) Wimbish Housing, Parsons + Whittley Architects. UK Passivhaus Awards 2012: residential category. www.passivhaustrust.org.uk/UserFiles/File/Projects/Awards2012/Wimbish_Parsons+ Whittley.pdf, London, accessed 20 September 2014.

Passivhaus Trust (2011) *Passivhaus and Zero Carbon: Technical Briefing Document*, Passivhaus Trust, London.

Price, S. and Brown, H. (2012) Viewpoint: mechanical ventilation with heat recovery – Designing and implementing a robust and effective ventilation system. Encraft. www.encraft.co.uk/wp-content/ uploads/2012/08/Viewpoint-August-2012-MVHR-Designing-and-implementing-a-robust-and-effective-ventilation-system3.pdf, accessed 1 August 2012.

Reason, L. and Clarke, A. (2008) *Projecting Energy Use and CO2 Emissions from Low Energy Buildings: A Comparison of the Passivhaus Planning Package (PHPP) and SAP*, The Association for Environment Conscious Building, Llandysul.

Richards, J. (2013) Does the Passivhaus Planning Package make adequate allowance for accepted occupancy practices in the UK? MSc thesis, Centre for Alternative Technology, Machynlleth, Powys, UK.

Schiano-Phan, R., Ford, B., Gillot, M. and Rodrigues, L.T. (2008) The Passivhaus standard in the UK: Is it desirable? Is it achievable? *PLEA 2008 – 25th Conference on Passive and Low Energy Architecture*, Dublin, 22–24 October.

Schnieders, J. and Grove-Smith, J. (2013) Passive Houses in tropical climates. Passipedia. www. passipedia.org/passipedia_en/basics/passive_houses_in_different_climates/passive_house_in_ tropical_climates, accessed 16 December 2014.

Schnieders, J., Feist, W., Schulz, T., Krick, B., Rongen, L. and Wirtz, R. (2012) *Passive Houses for Different Climate Zones*, Passive House Institute, Darmstadt.

Schoenefeldt, H. (2014) *Interrogating the Technical, Economic and Cultural Challenges of Delivering the PassivHaus Standard in the UK*, Centre for Architecture and Sustainable Environment, University of Kent, Canterbury. https://kar.kent.ac.uk/44559/1/PassivHaus_UK_eBook.pdf, accessed 5 June 2015.

Shove, E. (2003) *Comfort, Cleanliness and Convenience: The Social Organization of Normality*, Berg, Oxford.

Shove, E. and Moezzi, M. (2002) What do standards standardize? *2002 ACEEE Summer Study on Energy Efficiency in Buildings*. Pacific Grove, California.

Simmonds Mills (2010) Sustainability awards 2010; UKGBC. http://simmondsmills.com/projects/files/ UKGBC_sept_02_final_submission.pdf, accessed 30 November 2014.

Taylor, M. and Raffiuddin, T. (2013) *How to Develop User Guidance for a Passivhaus Building*, The Passivhaus Trust, London.

Tebbutt, L. (2010) First past the post. *Grand Designs Magazine*, October, 58–63.

Tiramani, M. (2013) Y Foel, the UK's 1st certified Passivhaus dwelling. http://passivebuild.co.uk, accessed 17 June 2013.

Tofield, B. (2012) *Delivering a Low-energy Building: Making Quality Commonplace*, Adapt Low Carbon Group: Norwich.

7 Ventilating buildings

7.1 Introduction

Ventilation significantly influences the energy demand of a building. The dissipation of energy in the air that leaves a building can account for as much as 40 per cent of a building's total primary energy demand (Pérez-Lombard *et al.*, 2008). Indeed, the energy demand attributed to heating, ventilation and cooling typically accounts for more than half of total primary energy demand in developed countries, and it is the exchange of indoor air with the outdoor environment that has been cited as one of the main aspects driving that energy demand (Santos and Leal, 2012).

The UK Department for Communities and Local Government defines ventilation as:

> The supply and removal of air (by natural and/or mechanical means) to and from a space or spaces in a building. It normally comprises a combination of purpose-provided ventilation and infiltration.
>
> (DCLG, 2010, p. 8)

It is important to note that this definition includes both ventilation that is designed (termed 'purpose-provided') and that which is unwanted (termed 'infiltration'). Purpose-provided ventilation refers to air exchange that is controlled using a range of natural and/or mechanical devices (e.g. windows, doors and vents). Infiltration is 'the uncontrolled exchange of air between inside a building and outside through cracks, porosity and other unintentional openings in a building' (DCLG, 2010, p. 8). The air can be flowing in either direction but is likely to be flowing outwards in terms of heat losses, typically through cracks and gaps in the building envelope. Infiltration is also sometimes termed 'air leakage'.

Following the often-quoted mantra '*build tight, ventilate right*', building designers have sought to reduce heat loss by increasing insulation levels and improving the airtightness of buildings. This therefore reduces the infiltration rates and overall air change rates of our buildings (typically measured in air changes per hour (ach), see Section 2.4.5). By changing the airtightness levels of a building, the overall ventilation rates will change too; the higher the level of airtightness, the lower the ventilation rate. Consequently, as airtightness levels improve, traditional ventilation strategies (including a reliance on infiltration, purging air through opening windows and intermittent air flows from fans) will become increasingly inadequate to maintain air change rates and a healthy indoor environment.

As a result, ventilation has become progressively mechanised in new and retrofitted buildings. In particular, mechanical ventilation with heat recovery (MVHR) is increasingly being applied in new housing in order to satisfy the goals of minimising energy demand without compromising indoor air quality (Crump *et al.*, 2009). In the UK, the National House Building Council (NHBC) has estimated that one-quarter of all new homes are incorporating these systems (NHBC, 2013). Indeed, low-energy standards (e.g. the Passivhaus standard) actually specify air exchange rates so low that, in conjunction with stringent energy demand requirements, the specification of mechanical ventilation becomes almost expected when endeavouring to meet the design criteria. Moreover, in order to meet the challenging energy demand requirements, the inclusion of heat recovery within this approach to mechanical ventilation is similarly inevitable (see Chapter 6). For a building with openable windows that can provide natural ventilation in a temperate climate, it may not be necessary to operate the MVHR system during the summer months.

As a response to the drive towards low-energy buildings (and in particular housing), the adoption of MVHR as a design standard is one that may be counterproductive as it can result in unanticipated negative results (Roaf *et al.*, 2009). This is because both the reduced ventilation rates and the mechanisation of ventilation can have undesirable impacts for indoor air quality and mould contamination (Laverge *et al.*, 2011). Indeed, concerns about increased risks of health problems, in particular respiratory health, related to inadequate ventilation in energy-efficient buildings, have been raised (Crump *et al.*, 2009).

In addition, the mechanisation of ventilation has an energy cost. While we know a great deal about the technical efficacy of such systems when they are operated in optimal contexts (i.e. factory testing in controlled environments), in reality there may be a counterproductive increase in energy demand that can erode a proportion of energy-efficiency savings made, and potentially increase carbon emissions. The balance of energy cost and saving depends on how the systems are installed and operated.

This chapter explores why ventilation is important, focusing on 'purpose-provided' ventilation rather than unwanted ventilation that occurs through 'infiltration' (which is discussed in Section 2.4.5). This is followed by an examination of: different types of ventilation, both mechanical and passive; the energy and carbon consequences of mechanised ventilation; and the challenges faced by occupants operating MVHR.

7.2 Why is ventilation important?

As we have discussed, one of the principal and critical elements of an energy-efficient building is the minimisation of unwanted infiltration and exfiltration of air, which reduces the amount of heat lost through draughts. Exfiltration is the leakage of air out of a space, whether intentional or otherwise. Efforts by industry and policy-makers to reduce unwanted infiltration have resulted in ever lower air change rates being sought for both new-build and retrofitted low-energy buildings. Despite this pursuit of significantly reduced air leakage, it is essential to keep in mind that a certain level of replacement of air is necessary, not least of all to breathe (hence no building will ever truly be 'airtight'). Indeed, we have all experienced the 'sleepy' feeling that arises in a crowded room, as the oxygen is replaced with carbon dioxide.

Therefore, to ensure a healthy indoor environment, ventilation in buildings has a number of critical roles (CIBSE, 2005), including:

- To provide adequate background movement of air through the building and thereby maintain indoor air quality by removing/diluting pollutants in occupied spaces, so as to ensure occupant health remains unaffected.
- To provide a heat exchange mechanism (heating and cooling), allowing for the thermal energy in outgoing air to be efficiently transferred to cooler incoming air.
- To control humidity and moisture.
- To prevent condensation and mould growth.
- To provide natural cooling, and overall thermal comfort through air movement (e.g. in warm climates).
- To provide adequate ventilation for effective operation of equipment and processes (e.g. to prevent a computer data centre from overheating).
- To exhaust pollutants from localised areas and processes.

A central aim of a ventilation strategy is therefore to provide an appropriate level of indoor air quality by limiting the accumulation of moisture and indoor air pollutants, which would otherwise become a health risk (CIBSE, 2005).

Indoor air pollution is defined as chemical, biological and physical contamination of indoor air (OECD, 1997). Contaminants include: organic compounds; inorganic gases; particulates; radon gas; and moisture. We now discuss the background of each of these contaminants in turn:

- *Organic compounds*, such as volatile organic compounds (VOCs), phthalates and formaldehyde, are emitted as gases from a wide variety of materials found in buildings. The list of sources is vast, including: building materials; particle boards; paints; varnishes and other coatings; furniture; textiles; carpets; plastics; consumer electronics; cleaning products; personal care products; air fresheners; tobacco smoke; insecticides; and pesticides (Mendell, 2007). Levels of VOCs have been found to be highest in new buildings and are associated with off-gassing from building products and internal finishes and furniture (Bone *et al.*, 2010). Chemical pollutants, such as VOCs in materials, can be radically reduced or eradicated from source during production by the removal or lowering of certain chemical pollutants, whether by regulation or voluntary agreements.
- *Inorganic gases* include carbon dioxide (CO_2), carbon monoxide (CO), nitrogen dioxide (NO_2) and ozone. CO and NO_2 arise from the combustion of fuels for heating, cooking appliances, and transport and industry. CO_2 in the indoor air quality context is produced mainly by the exhalation of humans and pets. Ozone is produced by electrical equipment (e.g. photocopiers) and from polluted outside air entering the building.
- *Particulates* (or particulate matter) refers to airborne particles, including PM_{10} and $PM_{2.5}$, derived from biological and non-biological sources. Biological sources include

those arising from: occupants, tobacco smoke, pets and their pests, house dust mite faeces, fungal spores, parts from moulds, yeasts, bacteria and pollens. Other particulates are derived from non-biological sources that enter the home in 'fresh' outside air and are derived from sources such as agriculture, industry and transport. Particulates pose a threat to human health because they are small enough to be breathed in and travel deep into the lungs. In particular, $PM_{2.5}$ arising from smoking and other sources are linked to chronic obstructive pulmonary disease (COPD) (Osman *et al.*, 2007).

- *Radon gas* is a naturally occurring radioactive gas produced by the radioactive decay of elements found in rocks and soils, and is a particular issue where natural levels of radon are high (e.g. the south-west UK). Radon seeps into buildings through gaps in the built fabric and accumulates unless ventilated out. Radon has been attributed as the second leading cause of lung cancer, second only to smoking (Zeeb and Shannoun, 2009; Darby *et al.*, 2005). Increasing the thermal efficiency of buildings has a complex effect on the accumulation of radon gas in buildings. This is because increasing the airtightness and insulation levels of a building both reduces the opportunities for admission of radon gas, thereby reducing the amount of gas entering a building, but also increases the rate of accumulation through reducing the opportunities for radon gas to leave a building.

- *Humidity* arising from activities such as cooking, washing and drying of laundry indoors causes condensation, damp and mould growth. Humidity is one of the main issues in housing, but is less of a concern in other buildings. The relationship between damp and mould growth and respiratory problems is well established, particularly wheezing and asthma in children (Venn *et al.*, 2003; Fisk *et al.*, 2007).

The creation of indoor air pollutants is inevitable and cannot be eliminated, but sufficient ventilation can ensure an appropriate air quality and humidity level is maintained. But what is an adequate rate for ventilation? This is variable and dependent upon the use of the building and space (CIBSE, 2005), with recommendations ranging from 0.5 to 40.0 ach (Table 7.1). A ventilation rate of 0.5–1.0 ach is considered to be adequate for homes (CIBSE, 2005). There is a well-documented association between ventilation rates below 0.5 ach and effects on occupant health, in particular respiratory ill health (Bone *et al.*, 2010).

Table 7.1 Recommended air changes per hour (ach) rates for different building types

Space use category	Air changes per hour (ach)
Public and commercial buildings (general use)	3.0–9.0
Homes	0.5–1.0
Assembly halls and auditoria	6.0–10.0
Hospitals and health care buildings	6.0–10.0
Laboratories	6.0–15.0
Catering	30.0–40.0

Source: CIBSE (2005).

Sick building syndrome (SBS) is a term coined to describe a building in which a high incidence of a range of symptoms, such as low levels of occupant perceived comfort, eye, nose and throat irritation, nausea, lethargy, headaches, respiratory problems, wheeziness and breathlessness, colds and infections, can be experienced by its occupants and which appear to be linked to time spent in the building (Finnegan *et al.*, 1984). A significant cause of SBS is inadequate ventilation and has been found to have particular associations with well-sealed buildings with mechanical ventilation or air conditioning (Wargocki *et al.*, 2002).

More recently, a relationship between increasing energy efficiency in buildings and an increased risk of health impacts has been identified. A study of social housing households compared Standard Assessment Procedure (SAP) ratings, as a proxy for airtightness of properties, with incidents of asthma (Sharpe *et al.*, 2015). The study found that a one unit increase in SAP was associated with a 2 per cent increased risk of asthma, with the greatest risks in homes with a SAP rating greater than 71 (which corresponds with an Energy Performance Certificate rating of C and above). Other studies have linked low air change rates (specifically of less than 0.5 ach) with increase rates of asthma in children (Bornehag *et al.*, 2005; Hagerhed-Engman *et al.*, 2009).

7.3 Different types of ventilation

Strategies for providing adequate ventilation in buildings can be divided into two types: (1) natural ventilation, and (2) mechanical, or forced, ventilation. We now discuss each of these in turn, before focusing on mechanical ventilation for the remainder of this chapter.

7.3.1 Natural ventilation

Natural ventilation is the traditional ventilation strategy and, as it typically requires no energy to operate, is also often referred to as 'passive ventilation' (CIBSE, 2005). Natural ventilation exploits the pressure differential caused by air moving around a building and the buoyancy of warm air inside a building, which results in the movement of air through purpose-built openings such as windows, airbricks and vents. To be effective, all natural ventilation strategies require a source of fresh air, as well as a path through the building, for a controlled amount of air to move and to exhaust to the outside.

There are two types of natural ventilation strategies: (1) wind-driven and (2) buoyancy-driven (Figure 7.1a, b, c and d). We now discuss the background (and sub-classifications) of these two types:

- *Wind-driven* ventilation strategies, including single-sided or cross-flow ventilation, use the pressure differential that is generated by the wind to drive or force air through openings in the building (Figure 7.1a, b). In single-sided ventilation, openings are positioned on one side of a space and provide airflow into and out of that space. In cross-ventilation, openings are positioned on the windward side and the leeward side of a room or building, creating a flow of air.
- In *buoyancy-driven* ventilation strategies, including passive stack ventilation, the movement of air occurs owing to the differences in densities of air inside and outside a building, which is mainly due to differences in temperature (Figure 7.1c, d). In short, hot air rises so it makes sense to use it. Buoyancy-driven ventilation strategies rely on

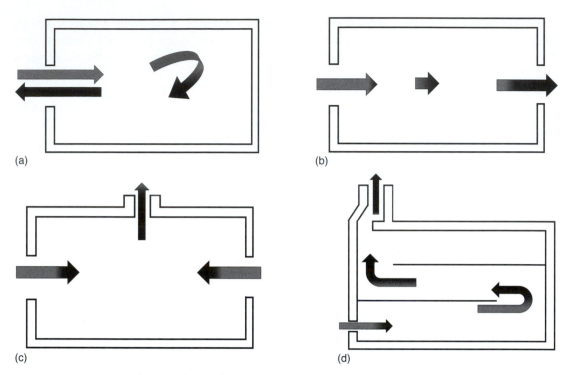

(a)   (b)

(c)   (d)

Figure 7.1 Diagram of natural ventilation strategies showing wind-driven and buoyancy-driven ventilation strategies: (a) single-sided ventilation; (b) cross-flow ventilation; (c) buoyancy-driven ventilation; (d) passive-stack ventilation.

the temperatures inside buildings being greater than those outside. Specifically, colder air enters the building at a low level and, as it is heated, it rises and exhausts from the building at a higher level, thereby 'sucking' cooler air in to replace it. Buoyancy-driven ventilation can also be assisted by stacks, wind towers, roof lights and the facade itself.

Natural ventilation systems are simple to operate, relatively low cost to install and (if entirely passive) do not require energy to operate. More complex natural ventilation systems can involve automation using, for instance, sophisticated control systems to operate openings and vents. In these automated cases, there will be an electricity demand.

Natural ventilation depends on the interaction between climate, design and use. Although natural ventilation offers a number of advantages, it also has a number of disadvantages (Table 7.2). Natural ventilation can be dependent on the weather (i.e. wind and humidity), local geography and surrounding environment, and as a consequence ventilation rates can be highly variable (see case study 3). Natural ventilation systems also require careful design to ensure adequate air movement in both cross-ventilation and passive-stack strategies. The effectiveness of natural ventilation systems is also highly dependent upon how the building occupant uses them. As a consequence, they can be difficult to control and hence their performance is difficult to predict (DETR, 2002). Being difficult to predict can also make such systems difficult to model with much accuracy.

Table 7.2 Comparison of key characteristics of natural ventilation and mechanical ventilation

Key characteristic	Natural ventilation	Mechanical ventilation
Risk of draught	Requires careful attention to vent/opening sizing, positioning and how controlled	Controllable system is designed and integrated appropriately
Control	Requires occupants to have access to openable windows and assumes that they will be 'properly' operated (although can be automated)	Control can be provided at individual level, but is generally centralised and can be costly. They can be fully automated and controlled by BMS
Closeness of control	Close control over temperature and humidity is not possible	Close control over temperature and humidity is possible, but involves higher energy demand
Capital costs	Capital costs can be low, but may be influenced by complexity of design or the built form required to achieve effective ventilation	Capital costs influenced by amount of mechanisation and airtightness of building
Running costs	Very low maintenance, especially if no automation	Requires regular inspection and maintenance. Energy costs are dependent on fan power, system design, building airtightness and efficiency of heat recovery
Flexibility	Reliance on ability of air to flow may impact on internal space use and the future flexibility of spaces (e.g. partitioning of spaces)	Highly flexible
Predictability	Performance can be modelled, but in reality this is subject to high variability related to weather/wind	Performance is predictable, but requires appropriate commissioning and maintenance
Noise	Silent in operation, but external noise transmission into building could be a problem	External and fan noise issues
Operation in polluted environments	Filtration of incoming air is problematic due to a reliance on pressure to drive airflow	Filtration is possible and included in many systems. Increased maintenance, filter replacement and higher energy demand an issue
Winter ventilation	Requires careful design to avoid excessive airflow, associated heat loss and thermal comfort issues	No issues; frost protection mechanisms standard

Source: CIBSE (2005).

7.3.2 Mechanical ventilation

Mechanical (also termed 'forced') ventilation strategies use mechanical equipment (e.g. fans, air-handling unit) to inject air into, circulate air through and exhaust air from a building. All of these require electricity for operation.

There are three main methods of mechanical ventilation: (1) pressure; (2) vacuum; and (3) balanced (Figure 7.2):

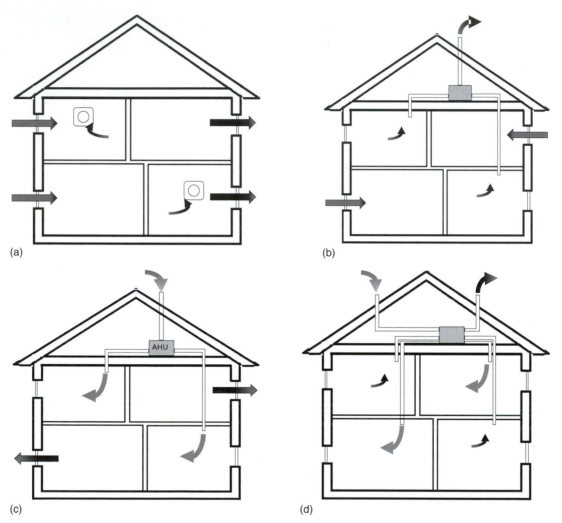

Figure 7.2 The main types of mechanical ventilation systems in residential buildings, showing (a) intermittent fan extract; (b) MEV pressure system; (c) MEV vacuum system; and (d) MVHR balanced system.

- A pressure system uses a fan or blower of some description to blow air into a building (Figure 7.2b), which raises the air pressure in the building to be slightly higher than outside.
- In a vacuum system, an exhaust fan blows air out of the building via a flue or stack. In this method the air pressure in the building is slightly lower than outside (Figure 7.2c).
- In a balanced system, the two methods (pressure and vacuum) are combined to create a balanced system. For example, in a home, fresh air is supplied to living and sleeping areas while exhausting stale air at an equal rate from bathrooms and kitchens (Figure 7.2d).

Mechanical systems can be as simple as intermittent fans (e.g. bathroom or kitchen extractor fans) (Figure 7.2a), or can be much more complex, including, for instance, continuous mechanical extract ventilation (MEV) (Figure 7.2b, c) or continuous balanced supply and extract with heat recovery (MVHR) (Figure 7.2d).

MEV provides continuous extract ventilation, which may be constant or variable, and may be controlled manually or automatically in response to specific physical criteria (e.g. levels of humidity or CO_2). Similarly to MEV, MVHR is a continuous extract system that uses a system of inlet and outlet ducting, but is used in conjunction with a heat exchanger that pre-heats fresh incoming air with heat extracted from the stale exhaust air. The heat recovery (HR) unit recovers heat that would otherwise be lost as the stale warmed air is exhausted from the building, thereby avoiding the need to provide (as much) heat, which helps to increase the energy efficiency of the building.

In contrast to natural ventilation systems, mechanical ventilation systems can be closely controlled. More complex mechanical ventilation systems, which are often combined with heating and cooling systems (also known as heating, ventilating, and air conditioning – HVAC) such as in large commercial buildings, can also be controlled by a building management system (BMS). In addition, MVHR systems only require inlet and outlet ducts and no other ventilation openings to puncture the built fabric. In this respect, mechanical ventilation systems can contribute towards achieving very high levels of airtightness.

Mechanical ventilation systems require regular inspection and maintenance of equipment to ensure optimal operation. For example filters require regular inspection, cleaning and/ or replacement to ensure they continue to let air flow and to perform their function of removing particulates from incoming fresh air (see Section 7.6.4)

7.4 Energy and carbon implications of mechanical ventilation with heat recovery systems

Mechanical ventilation systems require power to operate fans and other associated equipment (a significant contributor to what is termed parasitic energy demand) and therefore have an energy and carbon cost, with financial implications dependent upon the systems installed, which passive natural ventilation systems may not. This cost occurs both during occupation and is embodied while being manufactured.

7.4.1 Occupational energy and carbon of MVHRs

MVHR has a positive effect on reducing ventilation heat loss but a negative effect on power consumption (Laverge and Janssens, 2012). As a consequence, there is a trade-off between the mechanical ventilation and the heat recovery parts of the system.

The overall net energy balance will only be an energy saving if ventilation heat loss savings achieved by the heat recovery are larger than the power required by the fans in the mechanical ventilation system. If this is not achieved, the system as a whole will be a net energy consumer (Roulet *et al.*, 2001). The annual energy demand of mechanical ventilation systems is estimated to be 4–8 kWh/m^2 per year (Dodoo *et al.*, 2011; Tommerup and Svendsen, 2006). This energy demand has been limited by regulating minimum acceptable fan power (DCLG, 2011).

Heat recovery systems have been shown to significantly improve the energy efficiency of buildings (Zmeureanu and Yu Wu, 2007). Ventilation heat losses can be typically 35–40 kWh/(m^2·a) in residential buildings, with 80–90 per cent of this potentially being recovered (Tommerup *et al.*, 2007). A study on the simulated performance of MVHR systems in Finnish apartment buildings found that mechanical ventilation *without* heat recovery demands 67 per cent more energy in cold climates than when heat recovery is used (Jokisalo *et al.*, 2003). The inclusion of heat recovery has been estimated to result in a 20 per cent reduction in final energy consumption in homes in cold climates.

Studies have demonstrated this pattern in different countries. In an early study, Hekmat *et al.* (1986) compared different residential ventilation systems in different US climatic conditions, finding that the inclusion of heat recovery reduced the total heating energy demand by 9–21 per cent. More recently, a number of studies in Europe have been published, such as Maier *et al.* (2009), who compared the effect of ventilation systems in 22 low-energy homes in Germany. Their study found ventilation systems with a function of heat recovery have 10–30 per cent lower heating energy consumption than ventilation systems without heat recovery. This range is narrower than that found in an earlier European project, which found that MVHR systems reduced the total energy for space heating by 20–50 per cent, depending on climatic zone, building type and airtightness (Wouters *et al.*, 2001). These studies suggest that, while there is variation in the contribution that heat recovery can make in reducing heating energy demand, in most situations its contribution is sizeable.

MVHR systems have also been found to reduce carbon emissions. For example, Monahan and Powell (2011) found a 14 per cent reduction in the operational carbon emissions of homes using an MVHR system in conjunction with passive solar design strategies (for passive solar design see Section 2.4.6), when compared with three alternative carbon-reduction strategies which included solar thermal, PV and ground source heat pumps. In addition, the net carbon balance will also depend on the carbon intensity of the electricity supplied to power the system and the fuels used to provide the space heating. For example, Singh and Eames (2012) modelled the carbon implications of retrofitting a range of interventions in typical existing UK dwellings heated by gas central heating systems, including: reducing infiltration by 70 per cent; the use of MVHR and grid electricity; and MVHR with renewable generated electricity. The MVHR system reduced energy and consequently CO_2 emissions by 12 per cent. Critically, this increased carbon savings to 21 per cent when the electricity used was derived from renewables.

7.4.2 *Embodied carbon of MVHRs*

The inclusion of any new technology in the home has hidden energy and carbon burdens embodied during its manufacture, installation and end-of-life disposal (Chapter 3). Indeed, the specifying of MVHR will in the majority of cases result in the addition of a technology, rather than substituting for or removing the need for an existing one (the exception being those buildings in which conventional heating systems have been eradicated). This will result in greater material inputs and may also increase energy demand (Blom *et al.*, 2010). As a consequence, there will be a net increase in embodied carbon, which should be taken into account when considering a building's whole lifecycle.

There is a considerable lack of studies that quantify the embodied carbon impacts from technologies, such as MVHR. This is because quantifying the embodied carbon of an MVHR system over its lifetime is complex and requires many assumptions, not least on its production, design and installation, as well as its use and maintenance regimes. An early study by Nyman and Simonson (2005) estimated the embodied energy and carbon from the production and maintenance of an MVHR unit in Finland to be relatively small: 2,000 MJ of primary energy and 97 kgCO$_2$ per MVHR unit over a 50-year lifetime, not including end-of-life disposal. The study demonstrated that the energy demand and emissions resulting from electrical energy used by the fans far exceeded those of manufacture. Indeed, energy demand was 100–200 times greater in use than that required during production. The inclusion of heat recovery was found to increase embodied energy by 10 per cent, but was also compensated by recovering approximately five times as much energy as consumed by the fans. Nyman and Simonson (2005) expected similar conclusions in most locations, with the magnitude of benefits from heat recovery dependent upon local climate, energy sources, system efficiencies and ventilation rates. This suggests that the embodied energy burdens of production are relatively small and are compensated for by the net benefits of energy savings made during operational life.

Yet, while the amount of embodied energy is relatively small per unit, if MVHR becomes an industry norm for the majority of buildings, particularly new build homes and a significant number of energy retrofitted existing homes as suggested by the NHBC (2013), this relatively small amount will increase in significance. This could potentially offset a proportion of operational energy or carbon savings if these systems are not deployed effectively.

7.5 Technical challenges of operating MVHRs

There are many technical challenges in operating MVHRs that have an impact on the effectiveness and performance of these systems. These challenges include those related to the quality of the building itself and to how these systems are designed, installed and commissioned.

7.5.1 Infiltration and airtightness

The primary energy savings of MVHR systems are greater in low-energy buildings compared to conventional buildings, largely due to their higher levels of airtightness (Dodoo *et al.*, 2011). Unintentional infiltration of air can considerably reduce the performance of the MVHR system. This is because MVHR systems depend on balancing air supply and air extract. If the building envelope has a relatively high airtightness, then the two are automatically equal and therefore balanced. However, all buildings have infiltration to some degree and so differential flows may occur which will result in over ventilation, raising fan energy consumption and lowering the effectiveness of heat recovery. For this reason, there is a level of airtightness above which an MVHR system will become a net energy consumer, rather than saving energy. According to the Energy Saving Trust (2010), airtightness levels below 8 m^3/h·m^2 at 50 Pa (which is roughly 3–5 ach) is the point at which MVHR systems become more efficient (i.e. save more heating energy than they demand in electrical energy) than ventilation-only systems. Theoretically, the lowest potential energy use possible with an MVHR has been found to occur at an air permeability below 3 ach (Lowe, 2000). However, this is dependent upon the overall system efficiency, fan power required by the MVHR system and additional ventilation provided by window opening.

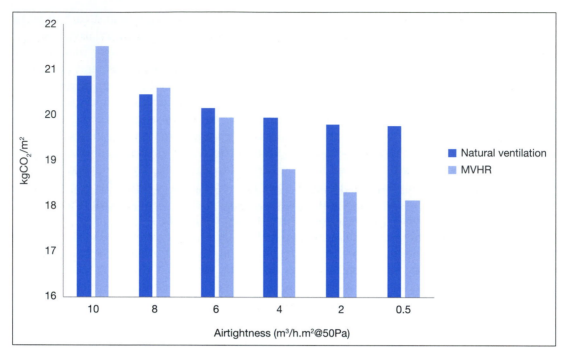

Figure 7.3 Modelled carbon emissions ($kgCO_2e/m^2$) for natural ventilation compared with MVHR for different airtightness rates.

Source: Monahan and Powell (2011).

The same is also true in terms of carbon. In a comparison of modelled carbon emissions of natural ventilation and MVHR in new-build UK homes, Monahan and Powell (2011) found there was no net carbon benefit until airtightness fell below 6 m³/h·m² at 50 Pa (Figure 7.3).

A study of energy demand in a typical building in the USA for different climatic conditions found that MVHR increased net energy demand as the parasitic energy generally outweighed the energy saved from heat recovery in typical homes (Sherman and Walker, 2007). Further studies of retrofitting MVHR in typical leaky UK homes support this, indicating that without significant reduction of infiltration and ventilation, there are no energy benefits or health benefits from such systems (Lowe *et al.*, 2000).

7.5.2 Design, installation and commissioning

Design, installation and commissioning can also significantly affect the performance of MVHR systems. Design problems include poor positioning of inflow and outflow vents and the specification of incompatible alternative ventilation strategies, including passive vents, mechanical extract vents and cooker hood extraction units (Laverge and Janssens, 2012). Installation problems include: missing insulation of ductwork and units; ductwork installed incorrectly; and supply and extract vents being inappropriately positioned (Lowe *et al.*, 2000). The Building Services Research and Information Association (BSRIA), a principal provider of building performance testing in the UK, found 95 per cent of homes tested failed to meet building regulations requirements. Reasons included ductwork incorrectly fitted (82.5 per cent of cases), missing or blocked with insulation and fans undersized or insufficient in number (Gilbert, 2012).

A recent review of evidence by the Zero Carbon Hub's Ventilation and Indoor Air Quality Task Group (Dengel and Swainson, 2013) noted that commissioning was also a common area of concern. There were significant differences noted between that recorded on commissioning certificates and the measurements made by researchers. Specifically, commissioning had not been carried out correctly and in adherence to available best practice advice from manufacturers and other bodies (e.g. NHBC, 2014). For example, some of the MVHR units inspected were running in boost mode constantly, proper controls were not always provided and the MVHR's filters were often clogged up with construction dust by the time the building was handed over to its occupants. Other commissioning problems commonly found have included fans operating at incorrect speeds and unbalanced flows, which lead to a 'two-fold' increase of electrical consumption (Lowe *et al.*, 2000; Laverge and Janssens, 2012). Dengel and Swainson (2013) also noted that competence within the industry is a key issue, in particular adherence to best practice, and they suggested that mandatory competency requirements for new-build ventilation systems could help solve this issue. A broader discussion of commissioning is provided in Section 4.9.4.

Ventilation systems are multifunctional in nature. In addition to providing fresh air, they also interrelate to heating, cooling and humidity. It is because of their multi-functionality that mechanical ventilation systems can be complex, particularly in non-residential or multi-occupancy residential buildings, and are, in such cases, rarely designed in isolation. The design and specification of such systems are generally the preserve of specialist building services engineers.

7.6 Occupant challenges associated with operating MVHRs

7.6.1 *Balancing purpose-provided ventilation by MVHR and window opening*

Purpose-provided ventilation using MVHR is typically applied in conjunction with openable windows. The two systems need to be balanced if the MVHR system is to operate effectively. Interestingly, post-occupation evaluation (POE) studies have shown occupants tend to open windows more often in low-energy homes with MVHR systems than in conventional homes without MVHRs (Stevenson and Rijal, 2008; Macintosh and Steemers, 2005). One suggested explanation for this is that occupants may have a preference for natural or demand control ventilation systems by convention and habit (i.e. it is what they have always done) (Laverge *et al.*, 2011). In further support of this, a case study of 22 homes in Germany found that occupants responded to being uncomfortable (including overheating, high levels of CO_2 and relative humidity) by opening windows, rather than by modifying the airflow rate of their MVHR systems (Maier *et al.*, 2009). Maier *et al.* (2009) suggested that this was because occupants had not yet developed a sense of the relationship between the length of ventilation time and indoor climate parameters that mechanised ventilation systems require. So, when activities occurred within the home that altered the indoor air quality in some way (e.g. increased temperature, CO_2 and humidity levels from having guests for a house party), occupants responded to the unpleasant indoor climate by opening windows rather than pre-empting it by altering MVHR ventilation rates beforehand. This inference is that opening windows for ventilation may be permanently related to habits which may be very difficult to overcome.

7.6.2 Operating MVHR controls

Studies of MVHR systems and 'passively adopting' occupants (i.e. people who do not actively choose a low-energy technology or building, but find themselves occupying a building containing such technologies) suggest that there may be a low level of interaction with the controls available to them (Lowe and Johnstone, 1997; Macintosh and Steemers, 2005). Macintosh and Steemers (2005), in a study of the Iroko Coin Street Development in London (UK), found that half of the development's occupants had made no adjustments to their MVHR system controls throughout the year. They suggested that the positioning of controls in inconvenient locations (e.g. putting controls in cupboards, corridors or above doors) was a contributing factor to inhibiting use. This emphasises the importance of the commissioner in setting up what may become the system's default setting in cases where the user never interacts with their control, preferring perhaps to leave the system running in the background as part of not taking an active role in its use (Behar and Chiu, 2013).

The design of controls may also be a key contributing factor impacting on behaviour (Stevenson and Rijal, 2008). Many flow rate controls, although presumably designed to be intuitive, consist of dials or digital screens showing four settings ('1', '2', '3', or 'boost'), with no indication of how these numbers relate to the rate of airflow and its effects on the efficacy of the system (Figure 7.4). The occupant has to interpret the meaning. However, if the occupant does not understand the system, they may not have the knowledge to interpret the meaning, leading to occupant uncertainty and an avoidance of interacting with the system. There is a need to relate the information and the design of controls to people's everyday lives (Foulds, 2013). In this case, rather than using the somewhat dry language relating to the control settings themselves, more relevant language could be used, such as 'I'm out' ('1'), 'normal living' ('2'), 'party time' ('3') and 'showering' or 'cooking' ('boost'), respectively.

7.6.3 Poor occupant understanding

Leading on from the previous discussion, poor occupant understanding is commonly talked about as being the root cause of occupants struggling to 'correctly' operate MVHR systems (e.g. knowing which control setting to use, or when the windows should be opened). Studies consistently cite low levels of occupant awareness and understanding of the purpose and functioning of MVHR systems, which many argue may be a key contributor to the poor management of the MVHR systems (e.g. Ezratty *et al.*, 2008; Macintosh and Steemers, 2005; Stevenson and Rijal, 2008). However, the evidence base is dominated by technical studies focusing on the operational efficiency of the technology and is predominantly based on models in which behaviour is reduced to quantitative data relating to, for example, how often and for how long a window has been opened, or whether a dial has been turned, or not, and to what setting is used in response to certain predefined environmental stimuli and its effect on the operational efficiency of the system.

Figure 7.4 An example of an MVHR controller showing settings 1, 2 and 3, with red LED signal light indicating when maintenance is due.

Source: J. Monahan.

A lack of, or inappropriate, interaction with the system components (e.g. controls and windows) by the occupants can lead to inefficient use, which is often perceived as a 'misuse' of the system. This misuse is frequently ascribed to occupant ignorance. This is in spite of there being little research on how passively adopting occupants actually understand or think about new technologies, such as MVHR. Under this model, the assumption is that 'good' understanding leads to 'good' use (i.e. information-deficit model). Nevertheless, while knowledge is clearly important, just 'knowing' what something is does not guarantee what 'action' is taken. Therefore, there is merit in activities that attempt to improve people's *practical* understanding of MVHR systems. However, the literature's traditional focus on the technical operation of MVHRs gives relatively little attention to how people practically use them, and often therefore focuses on the wrong kind of knowledge. To use an analogy: knowing what a bicycle is and what it does is very different to knowing how to practically use a bicycle, which requires tacit and sensory knowledge constructed over time and with experience. This said, even though it may be more likely, we are still keen to emphasise that there is still no guarantee that building occupants will use MVHRs 'correctly', even with a significantly enhanced practical understanding: there is a gap between knowledge and action, and we need to be more sensitive to this.

7.6.4 Maintenance and management of MVHR

Regular inspection and maintenance of MVHR systems (in particular, their filters) is essential to their efficient operation. Indeed, regular replacement or cleaning of filters is required to maintain the designed airflow rates and to minimise fan power and energy demand. Thus, poor maintenance results in soiled filters, which lower airflow rates and increase fan power and hence energy demand (Dengel and Swainson, 2013).

If MVHRs become increasingly prevalent in our building stock, as is looking likely, maintenance of MVHR systems will be critical to maintaining adequate indoor air quality. In particular, maintenance will help to prevent issues arising that are similar in nature to sick building syndrome, which were found to be prevalent in non-residential buildings in the 1980s and 1990s (Crump *et al.*, 2009).

Best practice guidelines recommend that the filters should be clean when the building is handed over to the occupant. In addition, a recommended maintenance schedule should be provided to the occupants at handover, so as to give them an idea of appropriate time intervals for maintaining their filters (NHBC, 2014). However, research suggests that maintenance does not routinely take place in new buildings (Dengel and Swainson, 2013). Furthermore, manufacturers of MVHR systems are reporting that there are few market opportunities for replacement filters, with some even reporting zero filter sales (Crump *et al.*, 2009). This therefore strongly suggests that maintenance may become a critical issue as uptake of these systems increases.

Maintenance may also be problematic for reasons of responsibility (Monahan, 2013). Responsibility for maintenance will vary depending upon the ownership arrangements of the home and MVHR equipment in it. In the case of rented properties, contractual arrangements will be required to ensure systems are maintained and this will have related costs for the lifetime of these systems.

We can speculate that it is likely that the majority of new homes will be privately owned, suggesting that there will be a reliance on the households themselves to institute maintenance regimes. Therefore, the householders themselves will be responsible for checking, cleaning and changing the filters regularly. Will these 'passively adopting' households take on the responsibility associated with this additional housekeeping task? This question and the issues discussed are explored in the first case study (Section 7.7.1).

7.7 Case studies

7.7.1 Case study 1: Lingwood, Norfolk, UK (Monahan, 2013)

Rationale

This case study considers the energy and carbon outcomes from the introduction of MVHR in new housing in the UK, where it is a relatively new technology. It builds on the issues posed in this chapter on the lifetime energy balance between embodied energy and energy demand.

Project background

- Location: Lingwood, Norfolk, UK
- Building type: residential
- Tenure: social rent and shared ownership
- Year of construction: 2008
- Size: a block of four terraced houses; floor area two 71 m^2 and two 83 m^2
- Construction approach: off-site closed-panel timber-framed, with additional insulation to floors, walls and roof, installed on-site and with larch cladding.

Approach

This interdisciplinary study explored the energy demand and carbon emissions of households that passively adopted MVHRs. First, the energy and carbon balance of the system in operation was estimated. Energy demand information (from one year of monitoring), SAP 9.81, and system manufacturers' data were used to model the theoretical energy demand of the mechanical ventilation. These were also used to quantify the energy savings from the recovered heat of the heat recovery element, at the tested airtightness of 6.25 m³/h·m² at 50 Pa. It was assumed that energy demand was met by grid-electricity and the heat recovery energy saved displaced gas.

In addition, the study involved two rounds of semi-structured interviews with each of the households (the first within the first month of occupation, the second at six months). The interviews ranged in duration between 30 and 60 minutes. The interviews were structured around topics including: information; understanding and knowledge; system control; and maintenance and ownership. In addition, informal observation of the occupants formed the basis of field notes. These qualitative data were collected to explore how the occupants understood and operated the MVHR systems.

Findings

The results of the energy and carbon balance illustrate the difference between end-use energy and primary energy (Table 7.3) for the three case study homes that operated their MVHR systems for the monitoring period. The average estimated energy demand suggests there was a net end-use energy saving of 295 kWh/yr. However, in primary energy terms there was a near balance in energy, showing a very small net increase in energy demand. There was also a net carbon increase of 28 $kgCO_2$/yr (Table 7.3) due to electricity having greater carbon intensity per kWh than gas.

The study found the overall energy balance to be negative (–2 kWh/year of primary energy) and carbon savings minimal (–28 $kgCO_2$/year). In these homes, the energy demands of the ventilation system were offset by the energy savings made by the heat recovery unit. This case study illustrates that installing MVHR systems on homes that fall above the indicative airtightness rate will not perform to optimum energy and carbon saving levels.

Table 7.3 Average estimated MVHR energy demand (kWh/yr) and carbon balance ($kgCO_2$e/yr) for a house

	Energy demand (kWh/yr)	Primary energy demand (kWh/yr)	Carbon dioxide emissions ($kgCO_2$ e/yr)
Parasitic energy demand (electric)	226	633	127
Heating energy saved (gas)	522	635	99
Net difference	–295	–2	28

Source: Monahan (2013).

None of the four households operated the system as expected by its designers, who assumed the controls would be operated to meet the changing ventilation rates in accordance to the activities and comfort of the household. Systems were disabled, used as if they were air conditioning or used intermittently, with households preferring to open windows. One household used the on/off switch to operate the system when they felt their home needed ventilation, rather than the four speed control to operate the system. Another household did not like the 'cold draughts' and switched the system off, never using it again. One household experimented with the settings and found the lowest setting to be most comfortable, which they then decided to keep it on constantly.

The information provided was generally thought of by the households as inadequate. It did not give the households the information they needed to operate their systems. In the absence of useable information, the households were left to build their own understanding and knowledge through experimentation and trial and error until they achieved the satisfactory outcome. This suggests that information provided to users of such technologies should be focused on the outcomes (or energy services) that the MVHRs are there to provide (e.g. comfort, good air quality) rather than on the technical working of the MVHR system itself.

For one household, the 'draughts' created from the system were associated with air conditioning that they had experienced on holidays and this framed their understanding and how they used the system. This emphasises the importance of past experience, as previously discussed in Section 7.6.1. During the summer, the 'cooling draughts' were perceived as beneficial. In this context the ventilation system was run on high or boost, with the doors and windows open to provide 'air conditioning'. However, during the winter the draughts were perceived as detrimental and the system was switched off. For this household, the MVHR system was clearly identified as, and used as, air conditioning for cooling and not related to the heating at all. Consequently, its operation was entirely contradictory to its designed purpose. If the mechanisation of fresh air becomes institutionalised as a standard residential technology and is identified by a significant number of households as a means of cooling, rather than associated with indoor air quality and heat (as suggested in the results of this study), then this could lead to an increased acceptance of air conditioning for space cooling in the home. Such a trend would be contrary to the intended outcomes of policy, leading to a load shift from heating to cooling and representing no energy demand reduction overall.

This study emphasises that MVHR systems, as currently deployed in new-build homes in the UK, may not meet their intended aims. Indeed, the net result may be an increase in the energy demand and carbon emissions of new homes, contrary to policy objectives.

Key messages

- MVHR systems may result in increased energy demand and carbon emissions if they are not designed, deployed and operated appropriately.
- The contributing influences of MVHRs demanding more energy than expected included:
  - o being installed in homes with sub-optimal airtightness;
  - o MVHR systems not being operated as intended, if indeed they are used by the occupants at all;
  - o no regular maintenance, cleaning or replacing of filters.

- The mechanisation of ventilation may lead to an increased acceptance of 'space conditioning' in buildings not traditionally 'conditioned', such as homes, and lead to the widespread acceptance of summertime space cooling, shifting or increasing summertime energy demand.

7.7.2 Case study 2: mechanical ventilation and indoor air quality in Dutch new-build housing (Boerstra et al., 2012)

Rationale

This case study is based on a Dutch national study into the performance of mechanical ventilation systems and its effect on health and perceived indoor environmental quality of occupants. The study illustrates the problems and issues of mechanical ventilation systems (both MEV and MVHR) in buildings and relates technical shortcomings with health outcomes discussed in Sections 7.2 and 7.5.

Project background

- Location: the Netherlands
- Building type: residential
- Tenure: mixed tenure (majority privately owned)
- Years of construction: 2006–2008
- Type: single-family homes consisting of terraced, semi-detached and detached homes
- Study sample of 299 homes (150 with MVHR and 149 with MEV).

Nearly all new-build dwellings in the Netherlands have mechanical ventilation, most commonly MEV or MVHR (Boerstra *et al.*, 2012). With increasing rates of installation, there is a notable increase in reports from occupants of new, energy-efficient Dutch homes (particularly those with MVHR) regarding noise, draughts, overheating, poor indoor air quality and lack of personal control (Balvers *et al.*, 2012). This has been reported widely on national television programmes and newspapers since 2008 and addressed in the Dutch Parliament.

Approach

Performance-related measures included ventilation rates on a per room basis, installation quality and noise levels. Furthermore, while no direct measures of indoor air quality indicators (e.g. CO_2, VOCs) were taken, characteristics indicative of health risk factors were used as proxy measures instead. These proxies included capacity of ventilation unit and vents, visual inspection of the ventilation units and ducts, and commissioning practices. Using all of these data, the performance of the ventilation systems was compared with the requirements in the Dutch Building Code (BRIS, 2003) and the Dutch GIW/ISSO guidelines for installations in dwellings set by the Dutch Guarantee Institute for Housebuilding (ISSO, 2008). The assumption was that a correctly installed and commissioned ventilation system would meet all requirements of both the Building Code and the GIW/ISSO guidelines.

Occupants were also questioned on their perceived indoor air quality and health. The survey contained standardised questions on non-specific health symptoms, indoor environment-related health symptoms and perceived indoor environment, sleep disturbance and the perceived controllability of their ventilation systems. Indoor environment-related health symptoms were determined by a selection of five items comparable to Burge *et al.*'s (1993) Building Symptom Index (BSI 5). These were:

- fatigue
- headache
- burning, itchy or irritated eyes
- irritated, stuffy or runny nose
- hoarse, dry throat.

Findings

Three main issues identified by the investigation included: insufficient ventilation rates; high noise levels; unclean systems and insufficient maintenance:

- Insufficient ventilation rates: overall <30 per cent of the homes were found to have ventilation rates that complied with minimum standards. Air supply rates were found to be insufficient in 85 per cent of homes with MVHR. Air exhaust rates were found to be equally poor – again 85 per cent of homes with MVHR and 76 per cent of homes with MEV. This was traced to technical issues that included incorrectly installed duct work in 48 per cent of cases. In addition, recirculation of exhaust air back into the air supply was found in 59 per cent of MVHR homes.
- High noise levels: the study concluded that noise was a significant issue in these homes. In 86 per cent of the houses, sound levels were higher than 30 dB(A). This was particularly pertinent when operating at the higher fan speeds. The researchers suggest that high noise levels could be caused by inadequate installation of the system or by its ductwork. Incorrect placement of the ductwork or sharp bends causes more air resistance, which makes the system work harder than necessary and so causes more noise. Maintenance is also a factor in keeping the system operating under the noise limit. For example, dust accumulation in ducts is a sound amplifier.
- Unclean systems and insufficient maintenance: the systems were also typically found to be poorly maintained. Regular maintenance was not common practice (66 per cent of MVHR and 82 per cent of MEV were not inspected at all or given any annual maintenance). Further, supply ducts were contaminated with dust and dirt in 77 per cent of cases. Air filters were visibly dirty in 43 per cent of MVHR homes.

Despite the issues found, the majority of householders perceived the indoor air quality as good or very good (66 per cent MVHR and 79 per cent MEV). However, indoor air quality was perceived less favourably in homes with MVHR when considering air quality, dryness or air, noise and control systems. Interestingly, there was no clear relationship between self-reported health and shortcomings of the ventilation, with only 5 per cent reporting a moderate increase in non-specific health symptoms. However, 17 per cent did report increased sleep disturbance relating to noise issues.

Three technical features (well-ventilated bedrooms, recirculation of air and high exhaust flow rate at night) of the ventilation systems were found to be significantly associated with perceived indoor environment or subjective health:

1   Lower non-specific health symptoms were found in homes with well-ventilated bedrooms.
2   There was a lower perceived air quality where recirculation of air occurred.
3   There was a greater noise annoyance in homes with higher measured overall exhaust flow rate at night.

In common with many other studies already cited in this chapter (e.g. Monahan, 2013; Stevenson and Rijal, 2008), most occupants did not use their ventilation systems as prescribed by the manufacturers. Indeed, in 96 per cent of the households studied, the control switch was mostly used in a lower control setting than is recommended for proper ventilation. This was found to be primarily related to noise levels and to overcomplicated switches. Many households under-ventilate their homes because of the high noise level disturbance of operating the MVHR at its highest setting. Households also explained that the MVHR controls and accompanying guidance information were too complicated.

Key messages

*   Many things can go wrong during the installation of mechanical ventilation systems of all types, including incorrectly designed and badly installed systems. For example, in this case study air ducts were not properly installed and air supply grills were not placed in optimal locations and recirculation of exhaust air was found to be occurring.
*   Maintenance is not always carried out as recommended. In this case study ventilation systems were not clean; internally dirty or dusty and air filters were dirty and required cleaning/replacement in 50 per cent of the homes. Further, for most of the households studied, there was no maintenance contract.
*   Noise is a critical issue. In this case study, it was the predominant quality issue with ramifications for under-ventilation and poor use of controls.
*   Despite the technical issues relating to design, installation and commissioning, perceived indoor air quality may not be a large issue in low-energy homes.

7.7.3 Case study 3: naturally ventilating buildings in the highly dense urban environment of Chinese cities (Tsou et al., 2012)

Rationale

This case study explores the implications of rapid urbanisation upon local wind environments, and considers what this means for sustainable building design decisions specifically relating to natural ventilation. We have chosen this case study to highlight the important role that the surrounding built environment plays in an individual building's ventilation strategy. It is also hoped that this case will implicitly emphasise that 'natural ventilation' is never truly 'natural', as it is so influenced by our surrounding infrastructure and landscape (which is by no means only 'natural').

Project background

- Location: Chinese cities
- Building type: relevant for all inner-city buildings
- Tenure: relevant for all tenures
- Year of construction: no buildings actually constructed, but the modelling was based on technological and city planning assumptions that were appropriate for 2012.

The rationale for this modelling study was that an increasing proportion of the global population are living in cities – as supported by the United Nations (2014), which states that 54 per cent of the world live in cities and that this will increase to 66 per cent by 2050. But, as our cities expand and become more densely packed, what does that mean for the wind environment within our urban areas and, consequently, the natural ventilation potential of our buildings?

Approach

Cities exist as part of a 'complex' and 'open' system, and are thereby affected by a huge array of factors within and beyond the city itself. Therefore the regional built environment and its climate needs consideration so as to better understand what exactly is required for a single building environment. Consequently, three interrelated scales were taken into account during modelling: (1) large scale, involving the regional transportation of atmospheric pollutants; (2) meso scale, involving local air quality and climate simulation; and (3) micro scale, involving urban landscape profiling and the consideration of wind tunnels. Such an approach allows for local wind conditions to be assessed and the air quality implications of naturally ventilated buildings to be understood. They also explicitly recommend that modelling efforts and one's understanding of planning decisions can be improved by actually visiting the site and gathering real-world observations and 'feelings' of the wind environment.

Findings

While said to be already highlighted by several other studies (e.g. that have considered cities such as Hong Kong and Singapore), this study argues that the density of high-rise buildings is also on a seemingly ever-increasing trend in Mainland Chinese cities. The consequence of this is that wind speeds are becoming slower and the wind is even being completely blocked in some areas. This inevitably has an impact on natural ventilation rates and flows in inner-city buildings.

As explained in the Approach sub-section of this case study, the researchers explicitly took a systems-level perspective that acknowledged the interconnections between one specific building and the contextual surroundings of that building. It can thus be inferred that managing these challenges is in large part an urban planning issue. Therefore, with the support of the Chinese Government's Ministry of Housing and Urban–Rural Development, a proactive urban planning approach was taken on this issue for the pilot low-carbon green city of Shashan, in Guangdong Province. Within this experimental city planning project, the city's blocks were not designed squarely, as is traditionally done in China.

Instead, the blocks were constructed with an inclined angle to the north, which complemented the prevailing easterly and south-easterly winds.

Such planning approaches are crucial because, as Ghiaus *et al.* (2006) highlight in a separate study, natural ventilation can be used for free-cooling (instead of air conditioning, for instance) during significant parts of the year, which can thereby reduce building energy demand. However, natural ventilation is not possible for many dense urban environments, which have lower wind speeds, higher temperatures, as well as noise and air pollution. In particular, we note their findings regarding 'street canyons' having considerably lower wind speeds than, for example, rural areas with undisturbed wind, which supports the findings of Tsou *et al.* (2012), and emphasises that natural ventilation can be unfeasible for many inner-city buildings. This is especially significant given that, as Tsou *et al.* (2012) point out, Chinese city planning authorities are not adequately considering this, and thereby are essentially committing city populations to some form of energy-demanding mechanised ventilation.

Key messages

- Many Mainland Chinese cities with densely packed high-rise buildings have very low wind speeds and, in some areas, no wind at all.
- Low wind speeds (in addition to higher temperature, noise and air pollution) can make natural ventilation unfeasible, which can lead to the installation of highly energy demanding mechanical ventilation systems or even air conditioning (for summer cooling).
- The Chinese low-carbon green city of Shashan (Guangdong Province) has successfully piloted an urban planning strategy to tackle this; blocks were not constructed squarely (as normal) and were instead angled to complement the direction of the prevailing winds.

7.8 Summary

Ventilation is the supply and removal of air into and out of a building, and is composed of both ventilation that is wanted (purpose-provided ventilation) and unwanted (infiltration). While airtightness is often discussed as a key means to achieving reductions in energy demand, it is important to remember that we require a certain rate of ventilation to maintain a healthy indoor environment. In addition, as buildings are increasingly becoming more and more airtight (perhaps for energy demand reasons), traditional ventilation strategies, which have relied on infiltration and purging through opening windows, have become increasingly inadequate in maintaining a healthy indoor environment.

In this chapter we have also discussed different ventilation strategies, and thereby compared and contrasted natural ventilation and mechanical (or forced) ventilation. For instance, from a designer's perspective, mechanical ventilation is more controllable and reliably quantifiable compared with natural ventilation strategies. We also make clear that ventilation has become increasingly mechanised as part of a strategy to achieve increasingly low air change rates, which has also led to an increased uptake in MVHR systems (due to their heat recovery capability allowing for enhanced energy efficiency). In addition to potentially reducing energy demand and lowering carbon emissions, MVHRs can help to provide good levels of ventilation, indoor air quality and thermal comfort.

However, as this chapter and its case studies have illustrated, there are also many issues and challenges. MVHR systems are frequently not installed correctly or operated appropriately, which has potential ramifications for indoor air quality and the health of building occupants. The research discussed here indicates that in addition to poor practice in designing, installing and commissioning these MVHR systems, the way in which they are operated is also proving to be a critical issue.

Further, if the mechanisation of fresh air becomes institutionalised as a standard residential technology, will this lead to an increased acceptance of air conditioning in the home? And will such an outcome result in an increase in summertime energy demand that could negate any efficiency gains made during the heating season? Such a trend would be contrary to the intended outcomes of policy, leading to a load shift from heating to cooling and representing no energy reduction overall.

Moreover, this trend may be exacerbated as overheating increasingly becomes an issue as global average temperatures increase. If so, there are doubts that MVHR systems will be able to achieve the required level of purging for effective cooling during these conditions. This 'failure' will be exacerbated by the operating conditions. Some building occupants may never use their systems, and some may just not use the controls available to them, while others may even want the technology to cool them. If overheating is not to become a significant health risk during summertime and air conditioning is to be avoided as a technological response, then passive ventilation measures in conjunction with good passive solar design (Section 2.4.6) may be necessary.

If these issues are not addressed, the widespread use of highly mechanised ventilation systems will not meet their intended aims. In reality, the actual net result may be an increase in health risks and an increase in the energy demand and carbon emissions of buildings, both of which are clearly contrary to policy objectives.

References

Balvers, J., Bogers, R., Jongeneel, R., van Kamp, I., Boerstra, A. and van Dijken, F. (2012) Mechanical ventilation in recently built Dutch homes: Technical shortcomings, possibilities for improvement, perceived indoor environment and health effects. *Architectural Science Review*, 55(1), 4–14.

Behar, C. and Chiu, L.F. (2013) Ventilation in energy efficient UK homes: A user experience of innovative technologies. University College London http://discovery.ucl.ac.uk/1403244, accessed 29 May 2015.

Blom, I., Itard, L. and Meijer, A. (2010) LCA-based environmental assessment of the use and maintenance of heating and ventilation systems in Dutch dwellings. *Building and Environment*, 45(11), 2362–2372.

Boerstra, A., Balvers, J., Bogers, R., Jongeneel, R. and van Dijken, F. (2012) Residential ventilation system performance: Outcomes of a field study in the Netherlands. *Proceedings of the 33rd AIVC Conference and 2nd TightVent Conference, Copenhagen, INIVE*.

Bone, A., Murray, V., Myers, I., Dengel, A. and Crump, D. (2010) Will drivers for home energy efficiency harm user health? *Perspectives in Public Health*, 130, 233–238.

Bornehag, C.-G., Sundell, J., Hägerhed-Engman, L. and Sigsgaard, T. (2005) Association between ventilation rates in 390 Swedish homes and allergic symptoms in children. *Indoor Air*, 15, 275–280.

BRIS (2003) Bouwbesluit (Dutch Building Code). www.bouwbesluitonline.nl (in Dutch), accessed 29 May 2015.

Burge, S., Robertson, A. and Hedge, A. (1993) The development of a questionnaire suitable for the surveillance of office buildings to assess the building symptom index: A measure of the sick building syndrome. *Indoor Air*, 93(1), 731–736.

CIBSE (2005) *Application Manual (AM) 10 Natural Ventilation in Non-domestic Buildings*, Chartered Institute of Building Services Engineers, London.

Crump, D., Dengel, A. and Swainson, M. (2009) Indoor air quality in highly energy efficient homes: A review. NHBC Foundation. www.zerocarbonhub.org/sites/default/files/resources/reports/Indoor_Air_Quality_in_Highly_Energy_Efficient_Homes_A_Review_NF18.pdf, accessed 29 May 2015.

Darby, S., Hill, D., Auvinen, A., Barros-Dios, J., Baysson, H., Bochicchio, F., Deo, H., Falk, R., Forastiere, F. and Hakama, M. (2005) Radon in homes and risk of lung cancer: Collaborative analysis of individual data from 13 European case-control studies. *British Medical Journal*, 330, 223.

DCLG (2011) *Domestic Building Services Compliance Guide*, 2010 edition, HMSO, London.

DCLG (2010) *Approved Document Part F: Means of Ventilation*, 2010 edition with 2011 amendments, HMSO, London.

Dengel, A. and Swainson, M. (2013) Assessment of MVHR systems and air quality in zero carbon homes. The National Housebuilders Federation Foundation. www.nhbcfoundation.org/Publications/Primary-Research/Assessment-of-MVHR-systems-and-air-quality-in-zero-carbon-homes-NF52, accessed 29 May 2015.

DETR (2002) Energy efficient ventilation in housing: A guide for specifiers on the requirements and options for ventilation. www.envirovent.com/images/uploads/files/GPG268-Energy-efficient-ventilation-in-dwellings%281%29.pdf, accessed 29 May 2015.

Dodoo, A., Gustavsson, L. and Sathre, R. (2011) Primary energy implications of ventilation heat recovery in residential buildings. *Energy and Buildings*, 43, 1566–1572.

EST (2010) *Sustainable Refurbishment: Towards an 80% Reduction in CO2 Emissions, Water Efficiency, Waste Reduction, and Climate Change Adaptation*, Energy Saving Trust, London.

Ezratty, V., Duburcq, A., Emery, C. and Lambrozo, J. (2008) Residential thermal comfort, weather-tightness and ventilation: Links with health in a European study (Lares). *Proceedings of the 5th Warwick Healthy Housing Conference*, Warwick University, UK.

Finnegan, M., Pickering, C. and Burge, P. (1984) The sick building syndrome: Prevalence studies. *British Medical Journal* (clinical research ed.), 289, 1573.

Fisk, W.J., Lei-Gomez, Q. and Mendell, M.J. (2007) Meta-analyses of the associations of respiratory health effects with dampness and mould in homes. *Indoor Air*, 17, 284–296.

Foulds, C. (2013) Practices and technological change: The unintended consequences of low energy dwelling design. University of East Anglia. https://ueaeprints.uea.ac.uk/48784/1/FOULDS_-_FINAL_PHD_THESIS_VERSION.pdf, accessed 29 May 2015.

Ghiaus, C., Allard, F., Santamouris, M., Georgakis, C. and Nicol, F. (2006) Urban environment influence on natural ventilation potential. *Building and Environment*, 41, 395–406.

Gilbert, A. (2012) Practical experience of common ventilation problems. In *Healthy Homes: Ventilation for Good Indoor Air Quality*, Good Homes Alliance, London.

Hägerhed-Engman, L., Sigsgaard, T., Samuelson, I., Sundell, J., Janson, S. and Bornehag, C.G. (2009) Low home ventilation rate in combination with mouldy odour from the building structure increase the risk for allergic symptoms in children. *Indoor Air*, 19, 184–192.

Hekmat, D., Feustel, H.E. and Modera, M.P. (1986) Impacts of ventilation strategies on energy consumption and indoor air quality in single-family residences. *Energy and Buildings*, 9, 239–251.

HM Government (2010) *The Building Regulations 2010 – Approved Document F: Ventilation*, Department for Communities & Local Government, London.

ISSO (2008) GIW/ISSO publicatie 2008 Ontwerp- en montageadviezen. Stichting ISSO en Stichting GIW, Rotterdam.

Jokisalo, J., Kurnitski, J., Vuolle, M. and Torkki, A. (2003) Performance of balanced ventilation with heat recovery in residential buildings in a cold climate. *International Journal of Ventilation*, 2, 223–236.

Laverge, J. and Janssens, A. (2012) Heat recovery ventilation operation traded off against natural and simple exhaust ventilation in Europe by primary energy factor, carbon dioxide emission, household consumer price and exergy. *Energy and Buildings*, 50, 315–323.

Laverge, J., van den Bossche, N., Heijmans, N. and Janssens, A. (2011) Energy saving potential and repercussions on indoor air quality of demand controlled residential ventilation strategies. *Building and Environment*, 46, 1497–1503.

Lowe, R.J. (2000) Ventilation strategy, energy use and CO_2 emissions in dwellings: A theoretical approach. *Building Services Engineering Research and Technology*, 21(3), 179–185.

Lowe, R. and Johnstone, D. (1997) A field trial of mechanical ventilation with heat recovery in local authority, low-rise housing. Centre for the Built Environment, Leeds Metropolitan University.

Lowe, R., Johnston, D., and Bell, M. (2000) Review of possible implications of an airtightness standard for new dwellings in the UK. *Building Services Engineering Research and Technology*, 21, 27–34.

Macintosh, A. and Steemers, K. (2005) Ventilation strategies for urban housing: Lessons from a PoE case study. *Building Research and Information*, 33(1), 17–31.

Maier, T., Krzaczek, M. and Tejchman, J. (2009) Comparison of physical performances of the ventilation systems in low-energy residential houses. *Energy and Buildings*, 41, 337–353.

Mendell, M.J. (2007) Indoor residential chemical emissions as risk factors for respiratory and allergic effects in children: a review. *Indoor Air*, 17, 259–277.

Monahan, J. (2013) Housing and carbon reduction: Can mainstream 'eco-housing' deliver on its low carbon promises? University of East Anglia. https://ueaeprints.uea.ac.uk/42363/1/2013MonahanJPhD.pdf, accessed 29 May 2015.

Monahan, J. and Powell, J.C. (2011) A comparison of the energy and carbon implications of new systems of energy provision in new build housing in the UK. *Energy Policy*, 39(1), 290–298.

NHBC (2014) Technical standards chapter 3.2: Mechanical ventilation with heat recovery. National Housebuilding Council, Milton Keynes.

NHBC (2013) Technical Extra October 2013. Issue 12. National Housebuilding Council, Milton Keynes.

Nyman, M. and Simonson, C.J. (2005) Life cycle assessment of residential ventilation units in a cold climate. *Building and Environment*, 40(1), 15–27.

OECD (1997) *Glossary of Environment Statistics, Studies in Methods*, United Nations, New York.

Osman, L.M., Douglas, J.G., Garden, C., Reglitz, K., Lyon, J., Gordon, S. and Ayres, J.G. (2007) Indoor air quality in homes of patients with chronic obstructive pulmonary disease. *American Journal of Respiratory and Critical Care Medicine*, 176, 465–472.

Pérez-Lombard, L., Ortiz, J. and Pout, C. (2008) A review on buildings energy consumption information. *Energy and Buildings*, 40(3), 394–398.

Roaf, S., Crichton, D. and Nicol, F. (2009) *Adapting Buildings and Cities for Climate Change*, 2nd edition, Architectural Press, Oxford.

Roulet, C.-A., Heidt, F., Foradini, F. and Pibiri, M.C. (2001) Real heat recovery with air handling units. *Energy and Buildings*, 33, 495–502.

Santos, H.R. and Leal, V.M. (2012) Energy vs. ventilation rate in buildings: A comprehensive scenario-based assessment in the European context. *Energy and Buildings*, 54, 111–121.

Sharpe, R.A., Thornton, C.R., Nikolaou, V. and Osborne, N.J. (2015) Higher energy efficient homes are associated with increased risk of doctor diagnosed asthma in a UK subpopulation. *Environment International*, 75, 234–244.

Sherman, M.H. and Walker, I.S. (2007) *Energy Impact of Residential Ventilation Norms in the United States*, Lawrence Berkeley National Laboratory, Berkeley, CA.

Singh, H. and Eames, P.C. (2012) Reducing the carbon footprint of existing UK dwellings: Three case studies. *International Journal of Environmental Studies*, 69(2), 253–272.

Stevenson, F. and Rijal, H. (2008) The Sigma Home: Towards an authentic evaluation of a prototype building. *25th Conference on Passive and Low Energy Architecture*, Dublin.

Tommerup, H. and Svendsen, S. (2006) Energy savings in Danish residential building stock. *Energy and Buildings*, 38, 618–626.

Tommerup, H., Rose, J. and Svendsen, S. (2007) Energy-efficient houses built according to the energy performance requirements introduced in Denmark in 2006. *Energy and Buildings*, 39, 1123–1130.

Tsou, J.Y., Chow, B. and Fu, W. (2012) Wind environment and natural ventilation simulation for sustainable building design in Hong Kong and other China cities. *Proceedings of the 14th International Conference on Computing in Civil and Building Engineering*, Moscow, Russia, 27–29 June. www.icccbe.ru/paper_long/0334paper_long.pdf, accessed 9 June 2015.

United Nations (2014) *World Urbanization Prospects: The 2014 Revision – Highlights*, UNDESA, New York.

Venn, A., Cooper, M., Antoniak, M., Laughlin, C., Britton, J. and Lewis, S. (2003) Effects of volatile organic compounds, damp, and other environmental exposures in the home on wheezing illness in children. *Thorax*, 58, 955–960.

Wargocki, P., Sundell, J., Bischof, W., Brundrett, G., Fanger, P. O., Gyntelberg, F., Hanssen, S., Harrison, P., Pickering, A. and Seppänen, O. (2002) Ventilation and health in non-industrial indoor environments: Report from a European Multidisciplinary Scientific Consensus Meeting (EUROVENT). *Indoor Air*, 12, 113–128.

Wouters, P., Delmotte, C., Faysse, J., Barles, P., Bulsing, P., Filleux, C., Hardegger, P., Blomsterberg, A., Pennycook, K. and Jackman, P. (2001) Towards improved mechanical performances of mechanical ventilation systems: Tip-Vent. http://cordis.europa.eu/documents/documentlibrary/54656411EN6.pdf, accessed 29 May 2015.

Zeeb, H. and Shannoun, F. (2009) *WHO Handbook on Indoor Radon: Public Perspectives*, World Health Organization, New York.

Zmeureanu, R. and Yu Wu, X. (2007) Energy and exergy performance of residential heating systems with separate mechanical ventilation. *Energy*, 32, 187–195.

8 Building futures

This book has examined the energy demand of buildings, and its carbon consequences. Reducing energy demand by improving the energy efficiency of buildings is key to many countries meeting their carbon emission targets. This is because it is seen as technically relatively straightforward compared with other sectors. Yet, in developed countries, despite significant improvements in their thermal efficiency, the energy demand of buildings remains stubbornly constant, or only declines slowly, due to the challenges posed by growing populations, the expectations of larger, more comfortable and better equipped living spaces, and an expanding commercial sector. Developing countries, with hopes of improving the living conditions of their populations, will add considerably to the increase in energy demand, as will their understandable aspiration to expand their commercial and industrial infrastructure.

8.1 Reducing energy demand: a stubborn problem

We have identified that reducing energy demand in buildings is fraught with challenges, especially as we move towards a carbon-constrained world. The interlinked challenges of fuel poverty, an increasingly ageing population, fuel security and an old, inefficient building stock all point to the need for an interdisciplinary approach when addressing these challenges.

The decarbonisation of the energy supply will play a significant role in reducing the environmental impacts of our energy demand, but the slow adoption of renewable energy, together with the challenges of nuclear power, results in this being a gradual process. It is clearly necessary to reduce our demand for energy, while at the same time reducing the environmental damage done by the energy we use. By focusing on demand we inherently tackle supply by reducing it. We cannot consume our way out of trouble by just shifting energy supplies. In practical terms, we need to both reduce our demand and decarbonise our supply if we are to have any hope of meeting our carbon targets.

Although there have been considerable improvements in the energy efficiency of buildings, the current rate of reduction in energy demand is unlikely to be fast enough if we are to meet our carbon targets. While the focus of emissions reduction tends to concentrate on meeting emissions targets by a few specific milestone years, what is important to remember is that greenhouse gases are long-lived, so it is the cumulative emissions that need to be tackled; the sooner they are reduced the better. Targets are, however, still useful to focus minds and political will and to identify the pathways for meeting those targets. This is particularly relevant for the building sector because, for example, to meet the UK 2050 80 per cent target, a near-zero-carbon building stock is needed in order to accommodate increases or, at least, a maintenance of the status quo in other sectors such as aviation (HM Government, 2011).

As we have seen throughout this book, this need for urgency to improve building performance and lower energy demand is difficult to achieve for many reasons. There is considerable inertia in the system, with very low levels of turnover of buildings, which often have a lifetime of over 100 years, even though this may not have been intended in the original design. Occupants can have a strong social attachment to buildings for cultural and historical reasons (Georgiadou *et al.*, 2012), making large-scale demolition and reconstruction unlikely. Also, as we have seen in Chapters 3 and 5, from an embodied carbon perspective, retrofit is often a less energy-demanding option. However, in numerous developed countries, the current policy focus tends to be on new build rather than retrofitting. As we have seen, a large-scale retrofitting programme is essential if we are to bring our existing building stock up to current standards. For example, in the UK, 80 per cent of buildings that will be standing in 2050 have already been built (Royal Academy of Engineering, 2010), which represents a huge number of buildings.

The journey through this book has explored the role of energy efficiency in reducing energy demand. Energy efficiency is predominantly concerned with technological solutions because it is concerned with improving the rate at which energy is demanded, without the energy user seeming to have to change *how* they actually go about demanding that energy (Chapter 2). Although only one priority of the energy hierarchy, energy efficiency tends to dominate initiatives targeting reductions in energy demand. This is partly because it is seen as more straightforward than energy conservation, which more directly involves changing people's behaviour. As we have seen, there are many influences underlying why people act in the way that they do, so pulling them in a less energy demanding direction can be fraught with difficulty.

Reducing energy demand does not, of course, just apply to the occupants of buildings during the occupation phrase of their lifecycle. A range of actors influence how a building is designed and constructed and thereby influence its operational energy demand. Moreover, as buildings become more energy efficient during occupation, the energy demand over the rest of their lifecycle becomes relatively more important. For example, significant energy is embodied in the additional materials needed for insulation (Chapter 3).

Improvements in the energy efficiency of buildings, however, do not always meet expectations, resulting in an energy performance gap in both new and retrofitted buildings (Chapter 4), whereby expectations do not match reality. This can be particularly the case when retrofitting our existing building stock and especially when dealing with improving hard-to-treat buildings. These buildings, which were not originally designed to be airtight, can also provide extra challenges when tackling air quality issues.

Air quality is important for low-energy buildings that are designed to be airtight, and which often require mechanical ventilation to ensure their air quality is maintained.

Mechanical ventilation, even with heat recovery, has an energy cost of its own, which in less efficient buildings can be excessive. But airtightness without a ventilation strategy can result in health issues, mould and damp (Chapter 7). The Passivhaus standard, which is gaining considerable support among many building professionals (Chapter 6), has very high levels of insulation to reduce the demand for energy for heating. However, even these buildings can have a higher than expected energy demand, although not to the same extent as the energy performance gap of non-low-energy buildings.

Part of the problem in constructing energy-efficient buildings may be the additional skills needed for their rather specialist construction, which highlights challenges associated with the skills gap and a need for training. A lack of a skills base in the construction industry is a barrier to increasing the speed at which energy-efficient buildings are constructed and existing buildings are retrofitted. The economic downturn has led to a decline in investment in training at a time when new skills associated with retrofitting and building energy-efficient homes are urgently needed. A lack of direction and certainty from governments can mean that the industry is unlikely to invest in the necessary skills training that is required for their workforce, at all levels, to help meet the challenge of constructing low-carbon buildings.

Energy efficiency is one of a set of major approaches that are currently being used to reduce energy demand. Although these approaches can result in significant improvements in reducing energy demand, this is not always the case and it is certainly not as simple and linear as is often assumed.

To draw together and reiterate the key messages of this book, as well as to explore potential ways forward, we ask the following three main questions:

1   What are the characteristics of the main approaches to reducing energy demand?
2   What will new directions in technologies mean for the future of buildings?
3   What can interdisciplinary, systemic thinking offer energy management?

8.2 What are the characteristics of the main approaches to reducing energy demand?

In this section we explore the common underpinnings of the main approaches used by governments and other institutions to define and tackle energy-related challenges.

When seeking ways to reduce energy demand in buildings, a single disciplinary approach, or at least one that just involves technical disciplines such as engineering, architecture and building science, is usually used. It is true that many technical solutions, such as the Passivhaus standard, can reduce the energy demand of buildings considerably. However, the energy savings can be less (and/or often different) than expected. Consideration of the social sciences is, usually at best, 'bolted on' so as to facilitate a particular technological solution, such as to ensure purchase/installation of a technology or so that people 'understand' how to use the technologies. This is often thought to enable the full potential of a technologically determined way forward.

However, a seemingly 'technological solution' is often social at its core, not just because that technology has a designer (who is a person, with their own preferences and ways of viewing the world), but also because the 'success' of that technology depends on building occupants and owners opting to take up and install the technology, and then use it in the 'correct' way.

Another characteristic of these dominant approaches to reducing energy demand is to focus on the 'users' of energy, often regarding them as barriers to achieving the desired objectives. What is generally meant by users are the occupants of a building (i.e. the 'end-users') who operate the respective building technologies on a day-to-day basis. However, these technologies have to be designed, constructed and maintained, all by different types of users. The energy demand of a building is obviously strongly influenced by the initial decisions made by the designer, not only in terms of matters like the thermal envelope of a building and its orientation, but also the choice of the systems used in the building, for example, heating and cooling. It is exactly in this way that energy management becomes not just one activity in its own right. It is contextual and dependent upon the different ways energy can be used in the built environment by different types of actors. An argument here is that everyone manages energy in lots of different ways, depending on what they are doing (e.g. working versus home life). It is, therefore, important to be specific contextually about the particular type of 'user' and 'use' of energy. The different energy management roles also highlight the interconnections that exist between these uses and users, because of how interrelated they all are and how they influence one another.

While there is a general appreciation of how fragmented the construction sector is, due to the involvement of multiple actors in many different roles operating on multiple sites, the sector is also highly interrelated. For example, everyone's work influences the same set of technologies (the building), which will together impact upon the lives and energy use of the occupants. Overcoming some of the challenges of retrofit projects, for example, will require improving the interconnections not only between occupants and contractors, but also between designers, supply chains and procurement teams. While many built environment professionals working on the same development may never meet, they are all connected through the 'visions' laid out by the client brief and designer's plans, albeit that their interpretations of those 'visions' may be rather different!

Another dominant assumption in trying to reduce energy demand is that people's use of technology is predictable and rational, and that people behave so as to optimise their personal 'utility' (i.e. maximise benefits, minimise costs – and these need not be just financial). In other words, it is assumed that they will work out rationally what to do, so that they can gain the maximum benefits. For example, if people consumed energy only on the basis of cost, then lights would always be turned off when leaving a room, showers would be preferred to baths and low-energy appliances would always be purchased. However, we are rarely as rational as this. For instance, households may be more interested in gaining different types of 'benefits' that on the surface do not seem rational, but do begin to make sense if considered in the context of social expectations. For example, all of the following may result in an increase in energy demand and costs: making your house feel more 'homely' (both to you and your guests) through 'adequate' lighting; a long, hot bath that is more relaxing than a shower; or a reluctance to spend more money on a low-energy appliance, even if it saves money in the long term. In addition, building occupants act within their personal constraints, so may have more pressing concerns than their energy bills, such as the demands of a young family or a struggling business.

The UK Green Deal (Section 5.5.2) is a good example of a scheme designed to overcome financial barriers to installing energy-efficient technologies and to provide access to capital. It is assumed that once these (financial) barriers are removed, householders will take a rational approach and install energy-efficient technologies. However, it proved unpopular as other concerns were at stake, such as inertia or not wanting to attach a debt to their homes. Even for householders who do take a rational approach, the relatively high interest rates of the Green Deal are not financially very attractive. Arguably, it was never going to work very well.

Another major assumption is that people fail to act in a rational way because they have insufficient information. Certainly, the need for better information flows can be identified right across the construction supply chain. The introduction of greater integration between the various actors involved and the information they produce (e.g. through integrated energy design and building information modelling (BIM) systems) should help to improve the flow of information, but this only addresses part of the problem. If people are provided with 'perfect' information, will they act any differently, especially if the intention is to change behaviour and reduce energy demand? The assumption in providing information is that people are keen to save energy, but just do not know how!

An example of information provision intended to change behaviour is the UK's DECC 2050 Pathways Calculator (DECC, 2013), which provides significant amounts of detailed information about different ways of reducing carbon emissions, by changing energy supply and reducing demand, in order to meet the 2050 target. The rationale for this tool seems to be that if we have sufficient information we will make rational choices about the energy we use and how we use it. That is, of course, assuming we have the time, incentive and ability to use the tool, and then the motivation to act on that knowledge. Also, the model does not include an indication of the likelihood of the carbon-reducing measures being adopted. For example, one of the energy-conservation measures, reducing the internal temperature of our homes to 16 °C, is very unlikely to happen given our current desire for increasing the warmth in our homes.

Leading on from this idea of rational behaviour is the belief that behaviour is predictable and that under specific circumstances people will behave in the same way; we have already seen that this is rarely the case. This not only applies to occupants, but also to the behaviour of professionals throughout design, construction and commissioning. The assumption of predictable behaviour also underlies many energy performance models with, for example, assumptions concerning the hours a building is occupied. Although it may be possible to accommodate a range of different assumptions about behaviour into a model, averages are generally used, and then surprise is expressed when these models and their averages fail to predict energy demand accurately.

There is also the common viewpoint that the provision of new energy-efficient technologies will not change how everyday life and thus energy demand is organised. It is assumed that people will use new technologies in the same way that they used their previous ones, but that is not generally the case; the way that energy is used in buildings does not always stay the same when new technologies are introduced. Rebound (Section 4.8.3) can have a considerable effect on energy demand in new or retrofit low-energy homes. This is even more the case in developing countries, where improved access to energy resources and improved efficiency enables households to afford to heat their homes to more comfortable levels. For example, less polluting and more energy-efficient lighting and cooking stoves will improve health and the quality of living while increasing energy demand.

8.3 What will new directions in technologies mean for the future of buildings?

In this section we explore the physical technologies that will structure the fabric of the building, as well as technologies used by occupants inside that built structure.

With changes in building regulations and a growing general interest in low-energy buildings, there is a small but significant move towards a simpler 'fabric first' approach to construction, which prioritises high levels of insulation and improved airtightness over 'eco-bling' (such as micro-generation technologies) or overly 'smart' controls. The resulting lower heat loss means these buildings usually require smaller-sized heating systems, often in conjunction with MVHRs. Although these buildings may include micro-generation, such as photovoltaic and solar thermal panels, the emphasis is predominantly on reducing demand through energy efficiency, prior to then decarbonising supply. One of the reasons for this is that it is cheaper to achieve the necessary standards by improving the fabric of a building than it is by providing renewable technologies. The lifecycle energy demand is also lower as, for example, less energy is embodied in the micro-generation and energy-supply technologies. The focus of 'fabric first' applies equally to both existing and new buildings, although retrofitting the existing building stock has additional challenges, especially when the existing structure is poor.

Energy-efficient appliances and lighting are also important. Their availability is driven in Europe by legislation encouraging energy-efficient design, such as the Eco-design Directive for energy-using products. There is also the possibility that the demand for energy-efficient appliances will increase if building heat loss rates improve so much (e.g. Passivhaus) that overheating becomes an issue.

The adoption of efficient appliances is usually reliant on developers, owners and/or occupants purchasing them, unless, as in the case of incandescent light bulbs, inefficient versions of products are phased out, obliging the purchase of more efficient products. However, this phasing out is generally a rather slow process due to the need to give the industry time to adapt.

One way forward for future buildings, which might avoid the difficulties that 'people' are thought to bring to reducing energy demand, is to make technologies easier to use and to increase the automation of energy systems using a 'smart homes' approach. Although much of the research focuses on 'homes', several of the points raised apply equally to other types of buildings. In addition, the smart homes literature tends to focus on electricity use, but the concept also encompasses heating, in particular heating controls. However, in the context of communication links with energy suppliers, many of the issues that smart homes address are underlined by the difficulties of balancing electricity supply and demand.

At its simplest, a smart home (or smart building) can improve controls and automate simple tasks such as using motion sensors to turn lights on and off. At the other end of the scale, remote connection and control (for occupants and their energy suppliers) of a building's heating, cooling and ventilation systems plus the appliances can be included. As discussed in Chapter 2, there are different aspects of a smart home: the smart meter, which provides one- or two-way information flow to the energy supplier; and the provision of improved controls and/or an energy display meter, which both aim to encourage the occupant to save energy. However, as discussed in Section 8.1, the conjecture that providing better controls will lead to a reduction in energy demand assumes that the lack of these controls is all that is stopping people from reducing their energy demand. In the same way, there is an assumption that occupants will respond rationally and appropriately to the feedback of a display meter, which as we have seen may not necessarily occur (Strengers, 2013).

There is a suggestion in the literature that a smart home may appeal to quite a narrow range of households (Taylor *et al.*, 2007). Given the difficulty that many occupants (not just householders) have with controls they are not familiar with, particularly if they are not intuitively designed, it does seem likely that this type of technology may appeal more to 'technophiles'. Occupants may also not focus on reducing energy demand, but may use improvements in the efficiency of their homes to improve levels of comfort.

In the future, it is likely that energy suppliers will increasingly use smart meters to reduce their peak electrical demand (potentially avoiding constructing new power stations) by directly 'managing' their customers' demand (i.e. moving or spreading the peak demand). For example, in return for favourable energy price tariffs, building occupants could allow energy suppliers to turn off their electricity supply to specific non-critical appliances (e.g. freezers) for short periods of time when there is a surge in electricity demand. This is already available in the UK for commercial and industrial users (Element Energy, 2012). However, this is likely to shift demand rather than reduce it. This may or may not have carbon emission consequences, depending on the electricity fuel mix at the time of the original peak demand and when the demand is being shifted to. Another approach is to offer customers 'time of day tariffs', which are cheaper when the demand is lower. In such cases it will be possible to 'time' the use of appliances, such as washing machines, so they are only used when the price is low. This could be manually controlled by smart ICT technology (e.g. smart mobile phone), or it could be automated.

It seems very likely that in the future there will be increased 'automation' of other parts of the building lifecycle. Further adoption of BIMs and other ICT solutions will automate some of the exchange of technical information within the construction supply chain. However, there is a danger in relying too much on the models that are usually involved. The representation of a complex system of a building in a simple model, which will then be used as a basis for future decisions, is fraught with difficulties. As identified in some of the case studies in Chapter 4 (regarding the energy performance gap), many of the problems associated with achieving a low-energy building stem from a lack of effective communication between the design team, contractors and suppliers, a lack of proper system commissioning and shortfalls of the handover process.

8.4 What can interdisciplinary, systemic thinking offer energy management?

In this section we explore alternative ways of thinking that could offer fresh insights into the future energy demand of our buildings and, in particular, potential ways to reduce it.

Throughout this book we have highlighted that the understanding of energy demand and the challenge of reducing it is highly complex, involving numerous different actors, approaches and disciplines. For example, the underlying influences behind the energy performance gap cross many disciplines. Poorly constructed buildings may appear to be a technical issue, but inadequate construction practices and the lack of a skilled workforce clearly have social dimensions.

Trying to reduce energy demand in the construction sector is difficult because the construction sector is complex, rather than just being complicated. In complicated systems, it is usually possible to work out solutions and implement them, but in a complex system the relationship between the cause and effect is less certain (Chapman, 2002). In this context, a system is taken to mean a set of elements linked together to make a complex whole (Chapman, 2002). A systems approach tries to deal with complexity by increasing the level of abstraction, meaning that it loses levels of detail, rather than dividing the problem into separate, smaller components that are often more straightforward to understand. This latter approach, called a reductionist approach, can be useful for many types of problems, particularly in the field of science, but has the disadvantage of losing how individual components (be they people, systems or institutions) relate to and interact with each other (Chapman, 2002). This is important because it is these interconnections that are often the reason for the complexity.

A reductionist approach was used by the Zero Carbon Hub (2013), with a linear process of design, procurement and construction being used to structure an investigation into the various problems associated with the energy performance gap. However, many of the problems actually stem from the interconnections between the different actors. Another interesting point to note regarding the Zero Carbon Hub's approach is that it stopped at the end of the construction process. It did not consider the complexity of how the occupants actually use the building. Even when examining the technical aspects of a building, it is often how the individual components, such as people or technologies, interact that can lead to problems rather than the components themselves.

However, it is imperative not to view reductionist and systems-based thinking as competing approaches. The two approaches can be complementary when addressing a difficult challenge (Chapman, 2002). For example, while we have shown in this book that a building may not perform as expected for a wide range of non-technical reasons, there is still the need to look in detail at the technical performance of a wall by, for example, using a reductionist approach.

To address the complexity of energy demand reduction, a systems-based approach is needed to take into consideration the interconnections: between different actors; between actors and the technical systems they interact with (such as between a building's technical control system, how it was designed and how occupants use them); and between actors and society. By society we are predominantly referring to the institutional, policy and social context within which the construction sector and the built environment operate.

This systems-based approach needs to embrace the whole lifecycle of a building, not just the construction, maintenance and eventual demolition/recycling of a physical building, but also the lifecycle as it relates to the planning, design, management and occupation of a building. This lifecycle approach can provide a useful framework for demonstrating the interconnections between the different actors and explore the different uses or ways of managing energy.

Another feature of a systems approach is the recognition of positive and negative feedback loops that link the components, thus increasing the complexity of the interconnections. We have seen in the context of the construction supply chain how the feedback loops between different teams often do not work well. For example, changes in the building design or the components purchased may not get fed back into the energy performance calculations.

Plsek (2001; as referenced in Chapman, 2002) provides an illuminating example of the difference between a linear and a systems approach by comparing them to throwing a rock and a live bird. When throwing a rock, its landing place can be calculated. However, this is far more difficult, if not impossible, when throwing a live bird. He goes on to say that one way forward would be to tie down the bird's wings and tie a rock to it before throwing! Although this would make calculating its landing place slightly easier, the capability of the bird would be lost. He thus proposes that a better way to predict the landing place would be to use a bird feeder! This illustration can be applied to constructing energy-efficient homes: the regulatory system forces the construction sector to build in a certain way (although providing few checks to ensure they do so), but like the bird, their capability to produce their own solutions is lost. Although the bird feeder solution is overly simplistic, and perhaps implies a rational approach that we have questioned, it may be that if, instead of regulations, the building sector is provided with the incentives and the right market conditions to determine their own solutions, these may well be more effective, in terms of energy demand and cost, and could stimulate innovation.

By using a systems approach to tackle the problems of retrofitting, a novel approach, involving numerous different actors and including financial institutions, has been developed in the Netherlands. This, we believe, goes some way to addressing some of the fundamental barriers to reducing energy demand, particularly with regard to getting the best out of industry.

8.5 One way forward: the Energiesprong concept

One potential solution to the challenge of reducing the energy demand of buildings, as well as addressing many of the challenges highlighted in this book, is a new concept developed by Energiesprong (a Dutch market development team) and initially funded by the Dutch government. Not only have they helped develop an innovative technological approach, but they have also addressed many of the other social and financial challenges we have identified, such as occupant buy-in, upfront costs, concerns about skills and quality of the workmanship, and the practical disruption of a whole-house retrofit (Energiesprong, 2014).

Energiesprong have negotiated an agreement between housing associations and construction companies to retrofit 111,000 social housing homes to net-zero-energy levels. In this context, this means that the homes will not use more energy (including heating, lighting, hot water and appliances) than they produce during the course of one year. The organisation worked with the housing associations to offer the construction industry the security of large-scale contracts if they could initially retrofit 2,000 properties to the following strict criteria (Transition Zero, 2014):

- *Energy performance guarantee:* the retrofit package comes with a 30–40-year energy performance warranty that is backed by an insurer.
- *One-week delivery*: the installation of the retrofit should only take one week. Currently the Dutch Energiesprong homes have a target of ten days or fewer, but plans are in place to reduce this. The occupants need to be able to continue living in their home for all, or the majority, of the time.
- *Affordability*: a business case is developed to finance the investment, whereby the energy cost savings largely pay for the investment. The target price is based on the net present value of the energy cost savings across the lifetime of the retrofit. In the Netherlands, the construction companies have achieved price reductions of 40 per cent compared to the first pilot projects constructed three years ago.
- *Attractiveness*: the retrofit package needs to improve the appearance of the home and improve the occupants' quality of life to help gain buy-in from the occupants

themselves. Indeed, it was considered that the products must be perceived as being desirable, fun and easy, which is not how retrofit products are normally marketed. In line with this, the Dutch tenants are also provided with new kitchens and bathrooms.

In the Netherlands, to meet these challenges, the construction companies developed an industrial-scale approach using off-site construction. The companies use three-dimensional scanning to measure each individual building. These measurements are fed into a BIM system to produce the technical drawings used to manufacture made-to-measure panels (including walls, windows, doors and the roof), which fit tightly together surrounding the existing building. Electricity is provided by high-efficiency photovoltaic panels on the roof, with heating provided by a heat pump and ventilation through mechanical ventilation with a heat recovery system. All the building services, such as the heat pumps and the ventilation systems, are housed in a small wooden shed in the garden. This then means that if access is required for maintenance or repair, the company will not need to disturb the tenant.

High-quality workmanship is ensured by the retrofit's 30-year warranty, which is backed by insurance companies. The security of both the large-scale, long-term contracts and the insurance back warranty, provides the construction companies with the collateral to raise low-interest finance. The financial savings from the reduced energy demand pays for the retrofit. On completion of the retrofit, instead of paying the utility companies for their energy, tenants pay a similar amount of money to the housing association for all their energy requirements. This creates a financial stream that pays for the retrofit.

The thermal energy demand is not capped, but for electricity demand the tenants receive a 'bundle' based on the household's historic electricity demand. Like a mobile phone plan, if the limits set by the bundle are not exceeded they are covered, but if the tenant's electricity demand exceeds the limits, there is an additional charge per kWh. After the retrofit, the homes are monitored and a visual display unit is provided.

This Energiesprong approach addresses several of the issues identified in this book, such as the lack of integration between the various actors in the construction sector and the tendency to focus on occupants rather than other actors in the supply chain. This approach to retrofit encourages different parties (e.g. construction companies, financial institutions, insurance companies) to agree to work together to develop the right market conditions for widespread uptake. The security provided by the long-term contracts helps to overcome difficult regulatory and financial conditions that have been identified as challenges to an often risk adverse construction sector, encouraging the building sector to work together to develop innovative solutions. For example, the need for construction companies to provide a warranty will encourage them to work with component suppliers to develop new and better components continually.

This programme predominantly focuses on the delivery of technical and financial solutions to reduce energy demand in existing buildings, with the assumption that energy will definitely be saved once the technology has been installed. However, buy-in from the occupants is actively encouraged by increasing the attractiveness of the homes. The homes are also monitored and if the expected energy demand is exceeded, it is investigated by the construction company and the housing association to determine whether the problem is technical or behavioural, so that the issues can be addressed (Transition Zero, 2014). However, there are currently no details about the extent of any technical or behavioural problems or how these issues will be addressed. It will thus be very interesting in the future to see the results of the monitoring of the Dutch scheme, including feedback from the tenants, and how well the scheme transfers to other countries.

The large-scale, long-term nature of the programme addresses the urgency of improving our current building stock. Also, by giving industry the freedom and financial security to find the best technical solutions it will lead to technical innovation not only by the construction companies, but also throughout the supply chain. This innovation, plus the requirement to meet specific standards, points to the need for high-quality workmanship, leading to investment in skills training at all levels.

While the programme focuses primarily on buildings during occupation, it also targets institutional organisations that influence the uptake of energy-efficient building technologies. The standardisation of the off-site construction process will inevitably also result in lower levels of waste and emissions during construction and should lead to improvements in end-of-life management. Although these buildings may be quite complicated to 'recycle', it seems highly likely that the innovative materials and construction methods developed will feed into the construction of new buildings in the future.

8.6 Final thoughts

In the first chapter of this book we highlighted the enormity of the challenge of reducing our energy use, with global energy supply more than doubling in the last 37 years. All sectors of the economy need to contribute to reducing energy demand, including the built environment. Buildings in both the residential and service sectors currently have an energy demand of over one-third of global energy and half of global electricity, more if the construction of buildings is included. The greenhouse gas emissions associated with buildings have more than doubled in the past 45 years, accounting for 25 per cent of energy-related global emissions in 2010, which are predicted to more than double again by 2050 based on the current policy regime (IPCC, 2014). If we are to successfully tackle the energy-related problems of climate change, fuel poverty and fuel security, the buildings sector needs to be a major contributor to reducing global energy demand.

Initially, the common stance of many policy-makers and academics was that it would be relatively straightforward to improve the energy efficiency of the new and existing building stock; however, as we have demonstrated throughout this book, it is proving far more difficult than expected. There is a particular challenge with retrofitting existing buildings. While many retrofit projects have successfully reduced energy demand, cost can be high and the energy savings are often less than anticipated. Although there are now a range of technologies that can help to reduce energy demand in buildings, how these technologies are designed, selected and used is preventing us from reaping their full benefits.

This is particularly concerning as we are really only at the beginning of the transition to widespread low-energy buildings. In terms of our existing building stock there is significant lock-in due to the long lifespans of buildings and their associated infrastructure. Although in many countries retrofitting is well established, this has so far mainly targeted the 'quick wins' – buildings that are relatively easy to improve, and those with households that are more engaged with reducing their energy demand. A similar story can also be told for the construction of new low-energy buildings. Although they involve fewer practical issues than retrofitting, they only form a small proportion of buildings currently being constructed, often for more forward-thinking organisations such as social housing providers and very engaged individuals. As demonstrated by some of the case studies, if these relatively niche projects face challenges then we are likely to confront significant obstacles during the wholesale transition to a low-carbon building stock.

This book has shown that reductions in energy demand can be realised if we continue with a mainly technical approach to the issue, but that we will be unable to make the necessary energy and carbon savings in the required timescale if we are to meet our emissions targets and the needs of a growing population with their higher aspirations. The current way of approaching this problem is just not working fast enough, and unless we have an injection of fresh ideas across disciplines the future that we are building will not be sustainable. We, however, believe that if we begin to take an interdisciplinary and systems perspective of energy use in buildings, significant and timely reductions in energy demand are achievable.

References

Chapman, J. (2002) *System Failure*, Demos, London. www.demos.co.uk, accessed 5 April 2015.

DECC (2013) DECC 2050 pathways calculator. www.gov.uk/2050-pathways-analysis, accessed 5 April 2015.

Element Energy (2012) Demand side response in the non-domestic sector, final report for Ofgem. De Montfort University, Leicester.

Energiesprong (2014) Energiesprong. http://energiesprong.nl/transitionzero, accessed 5 April 2015.

Georgiadou, M.C., Hacking, T. and Guthrie, P. (2012) A conceptual framework for future-proofing the energy performance of buildings. *Energy Policy*, 47, 145–155.

HM Government (2011) *The Carbon Plan: Delivering our Low Carbon Future*, The Stationery Office, London.

IPCC (2014) Technical summary. In: *Climate Change 2014: Mitigation of Climate Change. Contribution of Working Group III to the Fifth Assessment Report of the Intergovernmental Panel on Climate Change*, Cambridge University Press, Cambridge.

Plsek, P. (2001) Why won't the NHS do as it's told, plenary address. *NHS Conference*, July 2001.

Royal Academy of Engineering (2010) Engineering a low carbon built environment: The discipline of building engineering physics. www.raeng.org.uk/publications/reports/engineering-a-low-carbon-built-environment, accessed 30 May 2015.

Strengers, Y. (2013) *Smart Energy Technologies in Everyday Life: Smart Utopia?* Palgrave Macmillan, Basingstoke.

Taylor, A.S., Harper, R., Swan, L., Izadi, S., Sellen, A. and Perry, M. (2007) Homes that make us smart. *Personal and Ubiquitous Computing*, 11, 383–393.

Transition Zero (2014) White paper for the expansion of Energiesprong's approach to France and the UK. http://energiesprong.nl/wp-content/uploads/2014/07/Transition_zero-1.pdf, accessed 5 April 2015.

Zero Carbon Hub (2013) Closing the gap between design and as-built performance: Interim progress report. www.zerocarbonhub.org/sites/default/files/resources/reports/Closing_the_Gap_Bewteen_Design_and_As-Built_Performance_Interim_Report.pdf, accessed 30 May 2015.

Glossary

Affordable housing	Socially rented housing provided to households whose needs cannot be met by the wider market.
Air conditioning	The process of changing the properties (e.g. temperature, humidity) of air, to establish more comfortable conditions for building occupants.
Air leakage	The unintentional inflow of air into a building, typically through cracks and gaps in the building envelope.
Air quality	How polluted (or 'clean') the air that surrounds us is.
Airtightness	How resistant a building envelope is to inward or outward (unintentional) air leakage.
Behaviour change initiatives	Initiatives that seek to encourage individuals to change specific actions ('behaviours') so that they become, for instance, less energy demanding.
Benchmarking	To compare results (e.g. building performance) with other, typically averaged, external results.
BIM (building information modelling)	The generation of a digital representation of the physical and functional capabilities of a building, used to improve flows of information between team members.
BMS (building management system)	A system which monitors and controls a building's mechanical and electrical equipment, potentially including ventilation, lighting, heating, fire and security systems.
Buildability	Examining a building's design from the point of view of the construction team and people involved in the manufacture and installation of products and components.
Building	A physical structure, typically with walls and a roof, and usually intended for permanent use.
Building element	Physical parts of a building construction. For example, walls, floors, roofs and openings (e.g. windows and doors).

Building energy performance	How a building performs technically in terms of energy use.
Building regulations	Regulations governing how buildings are constructed.
Built environment	Physical constructions built by humans, including buildings, roads and urban spaces, the principal components being buildings.
Capital cost	A one-off fixed cost typically associated with the purchase of, for example, a product, service, building or land.
Carbon dioxide equivalent (CO_2e)	A universal unit of measurement used to indicate the global warming potential of a greenhouse gas, expressed in terms of the global warming potential of one unit of carbon dioxide.
Carbon footprint	The amount of greenhouse gas emissions produced by a specific activity or service.
Carbon sequestration	The storage of carbon that has been removed from the atmosphere, for example through photosynthesis, and stored in products manufactured from biogenic materials such as wood, straw or animal fibre.
Climate change (anthropogenic)	Changes in the climate caused by the anthropogenic emissions of greenhouse gases.
Coefficient of performance	The ratio of energy input to heat output. Used as a measure of the performance of heat pumps.
Consumer electronics	Appliances such as televisions and DVDs.
Delivered energy	See *Final energy demand*.
Direct greenhouse gas emissions	Carbon emissions from sources that are owned or controlled by the manufacturer or company of products or activities.
Direct rebound effect	Using money saved from energy conservation and efficiency measures to pay for more of the same energy service, such as using additional energy to warm a building to a higher temperature. See also *Indirect rebound effect*.
Embodied carbon	Embodied carbon represents carbon emissions (expressed as $kgCO_2$ or $kgCO_2e$) emitted as a result of primary energy use at each stage of a product's (e.g. building) lifecycle.
Embodied energy	Energy used in each lifecycle stage of a product or activity, including that used in obtaining raw materials, processing, manufacture, maintenance, repair and end-of-life disposal.
Emission factor	The amount of greenhouse gas emitted expressed as CO_2e relative to a unit of activity, for example $kgCO_2e$ per kg of material.

Energy	The ability to do work.
Energy conservation	Reducing energy use by using less of a product or service. For example, lowering the thermostat on a heating system.
Energy demand	The energy used to deliver a product or service.
Energy efficiency	Using less energy to deliver the same product or level of service. For example, installing LED lighting.
Energy efficiency index	The ratio between the energy consumption of a sold appliance compared to that of a reference appliance defined in the Eco-labelling Directive.
Energy hierarchy	A prioritised sequence of different actions for managing a (sustainable) energy system. Generally includes the following priorities: energy conservation; energy efficiency; renewables and low-carbon technologies.
Energy intensity	A measure of energy in terms of output such as GDP.
Energy management	The process of planning and operating energy supply- and demand-related technologies.
Energy Performance Certificate (EPC)	A certified rating of the notional energy efficiency of a building, from A (most efficient) to G (least efficient). Required in the UK whenever a property is constructed, sold or rented, to comply with the EU European Building Performance Directive.
Energy performance gap	The difference between the expected (typically modelled) and actual (measured) energy demand of a new building or energy-efficiency technology.
Energy security	Establishing and/or maintaining a reliable, affordable and accessible energy supply.
Energy services	Services obtained from energy resources, such as hot water or light.
Energy supply	The delivery of fuel to where energy is in demand, including energy used for extraction, transmission, distribution and storage.
EnerPHit	A certifiable energy efficiency standard for retrofitting existing buildings, which is intended as a cost-effective alternative to the Passivhaus standard.
European Union (EU-28)	A political-economic union of 28 Member States, most of which are located in Europe.
Final energy demand	Energy used by final end users after it has been transformed. Excludes losses in the energy system. Also termed *Delivered energy*.
Functional unit	Quantified performance of a product or activity used as a reference unit for lifecycle assessments. For buildings, this can be whole building, area, volume or heat loss per unit area.

Greenhouse gases (GHGs)	Gases in the atmosphere that absorb and emit radiation at specific wavelengths within the spectrum of infrared radiation emitted by the Earth's surface, atmosphere and clouds. Includes carbon dioxide (CO_2), methane (CH_4), nitrous oxide (N_2O), hydrofluorocarbons (HFCs), perfluorocarbons (PFCs) and sulphur hexafluoride (SF_6).
Handover	The process of handing over a building, usually between a contractor and the occupant, but can be between other parties. This should not just involve handing over keys on move-in day, but also encompass a much longer period of transition (perhaps of up to a year), within which new occupants are supported.
Hard-to-treat buildings	Buildings that meet one of the following criteria: no cavity walls; no loft space or access; or are off the gas network. Such criteria would make it difficult for buildings to have cavity wall insulation, loft insulation or an energy-efficient, gas-fired condensing boiler.
Heat exchanger/energy recovery unit	Technology used in heat transfer from one medium to another, often for the purpose of energy efficiency in heating and cooling. For example, air-to-air heat exchanger found in MVHR systems or ground-to-fluid heat exchanger found in ground source heat pump systems.
Heat loss	To maintain a comfortable temperature inside a building the output of any heating system must be equal to the heat lost from the building down a temperature gradient. Measured in watts (W).
Heat pump	A device that transfers heat from one medium to another for either heating or cooling, e.g. includes transfer of low-grade heat from the environment (i.e. air, water or ground) to higher grade heat in a building for space heating (i.e. via water (e.g. radiators) or air).
Heating load	The peak power of heating a building at one given moment in time.
Indirect emissions	From an energy-generation perspective, emissions that occur at a different geographic location to where the energy is used, e.g. electricity generated at a central power station. From a carbon footprint perspective emissions arising from the products or activities of the reporting organisation that are controlled or owned by another company.
Indirect rebound effect	Using the money saved from energy conservation and efficiency measures to pay for different services, such as driving further or flying abroad. See also *Direct rebound effect*.

Infiltration	See *Air leakage*.
Interdisciplinary	The integration of different disciplines, and their various perspectives, into one structured and coherent approach. This, therefore, cuts across the boundaries that separate contrasting disciplinary-based approaches.
Leadership in Energy and Environmental Design (LEED)	A rating system for sustainable buildings developed by the US Green Building Council. A LEED rating level is based on the number of points achieved across nine areas. It is a voluntary rating system applied globally.
Lifecycle assessment (LCA)	Compilation and evaluation of all input and output flows and potential environmental impacts of a system throughout its lifecycle.
Mechanical ventilation with heat recovery (MVHR)	A ventilation system with energy recovery via a heat exchanger.
Meter	A device that measures how much of something (e.g. gas, electricity, water) has been used.
Micro-generation	The production of heat or electricity on a small and typically local scale (e.g. solar panels on a building).
Natural ventilation	Natural ventilation exploits the pressure differential caused by air moving around a building and the buoyancy of warm air inside a building, which results in the movement of air through purpose-built openings such as windows, airbricks and vents.
Occupation	The period of time in which a building is occupied and its technologies used.
Off-site manufacture	The part of the production process of a building that occurs away from the building site, under factory conditions.
Organisation for Economic Co-operation and Development (OECD)	An organisation created to stimulate global economic development and world trade. Currently has 34 members.
Parasitic load	The electricity demand of a device when it is turned off (e.g. overnight) but still uses small amounts of energy. Known as standby power for electrical appliances.
Passive ventilation	See *Natural ventilation*.
Passivhaus	A voluntary energy-efficiency standard, originating in Germany in the early 1990s, which achieves a high level of heating performance through super-insulation and airtightness.
Post-occupancy evaluation (POE)	An assessment of the performance (not only energy-related) of a building after it has been handed over to the occupants, typically involving both the design team and occupants, in addition to quantitative monitoring of the building.

Practices	Routine behaviour that forms activities (e.g. cooking, showering) that have evolved over time. Termed 'theories of practice' by academics.
Primary energy	Energy in its original state prior to transformation. The total fuel used to generate heat and power.
Purpose-provided ventilation	Controllable air exchange by means of a range of natural and/or mechanical devices (e.g. windows, doors and vents).
Rebound effect	See *Direct rebound effect* and *Indirect rebound effect*.
Renewable energy	Energy derived from a source that is not depleted when used. Includes wind, solar, hydro and biomass.
Residential sector	Consists of all houses, flats and mobile homes, whether they are unoccupied or occupied, and rented or owned. This does not include other institutional buildings that people may also be living in (e.g. schools, barracks).
Retrofit (also termed: renovate, refurbish, weatherisation)	The process of modifying an existing building and its services to improve its thermal performance.
Services (sector)	A sector of the economy that includes buildings related to retail, health, education, government and entertainment.
Shared ownership	A UK system provided by housing associations and local authorities, whereby the occupier buys 25–75 per cent of a dwelling and pays rent on the remainder.
Soft Landings	An initiative to improve the performance of a building by improving communication and feedback within a construction supply chain and promote the continued involvement of the contractors for the building beyond the handover period.
Solar gain	The increase in temperature in a building that is a result of incoming short wave solar radiation through glazed areas such as windows and doors.
Solar photovoltaic (PV)	A device that utilises the photoelectric effect in semiconductor materials to convert solar energy directly into electricity.
Solar thermal	The use of sunlight to heat or pre-heat water, usually for the hot water supply in buildings. Uses daylight so produces heat even in cloudy conditions.
Split incentives	Occurs when the costs and benefits of a transaction do not accrue to the same person, e.g. the cost of thermal improvements to a building are paid for by a landlord but the benefits of lower bills are gained by the tenant.
Standard Assessment Procedure (SAP)	The methodology used in the UK to measure the energy rating of residential buildings, which calculates annual energy demand, energy costs and CO_2 emissions. As the

	basis for demonstrating compliance with building regulations, it provides a SAP rating of 1 to a 100, with a higher score representing a higher thermal efficiency.
Standby	The electrical power a device consumes when it is turned off but is plugged into an electricity supply.
Systems thinking	A holistic approach that takes into consideration interconnections between different components of a system, such as people, the technical systems they interact with and society.
Tenure	The, usually contractual, status of how a building is being occupied (e.g. owner occupied, privately rented, socially rented).
Thermal bridge (also known as a cold bridge)	A localised area that has a higher rate of heat transfer than the surrounding area caused by a break or discontinuity or penetration of the insulation, e.g. junctions between walls and floors, roofs and walls and around windows and doors.
Thermal conductivity	Standardised measure of how easily heat flows through a material (units $W/m \cdot K$).
Thermal envelope	The physical separation between the internal and external environment. The wall, roofs and floors form the thermal barrier.
Thermal insulation	A low-conductivity material used to inhibit either the loss or entrance of heat (be it by convection, conduction or radiation).
Thermal resistance	Measure of how much heat loss is reduced through a given thickness of a material ($m^2 \cdot K/W$).
Tonne km	Standard unit of goods moved, calculated by multiplying the load (in tonnes) by the distance it travels (in kilometres).
Total primary energy supply (or resources) (TPES)	The energy used within a country, including indigenous production plus imports, less exports and international marine and aviation bunkers. 'Bunkers' refer to energy delivered to shipping and aviation engaged in international aviation and navigation. TPES is measured in terms of million tonnes of oil equivalent (mtoe).
U-value	A measure of heat loss ($W/(m^2 \cdot K)$) typically used for assessing the performance of building elements. It details the amount of heat lost (W) for each square metre (m^2) of the material being investigated, when the temperature (K) outside is at least one degree lower than inside.
Ventilation	The mechanical and/or natural supply and removal of air between spaces in a building.
Watt	A measure of power; one joule per second.

Useful resources

Note: All websites were correct at time of publication.

Databases, statistics and information

Buildings energy data book (US DOE)	Database of energy consumption in US residential and commercial buildings http://buildingsdatabook.eren.doe.gov
Building Performance Institute Europe (BPIE)	European database for the energy performance of buildings www.buildingsdata.eu
Centre for Energy Epidemiology Data Service	Energy demand datasets including social science data www.energy-epidemiology.info/data
Defra Greenhouse Gas Conversion Factor Repository	Database of UK GHG emission factors www.ukconversionfactorscarbonsmart.co.uk
Department of Energy and Climate Change (UK)	UK statistics of energy, fuel poverty, emissions and climate change www.gov.uk/government/organisations/department-of-energy-climate-change/about/statistics UK energy policies www.gov.uk/government/policies?departments%5B%5D=department-of-energy-climate-change
Digest of UK energy statistics (DUKES)	UK energy statistics www.gov.uk/government/statistics/digest-of-united-kingdom-energy-statistics-dukes-2014-printed-version
Greenhouse gas conversion factory repository	Online tool and repository for greenhouse gas conversion factors for the UK for reporting of emissions www.ukconversionfactorscarbonsmart.co.uk

HEED	Database of energy efficiency measures in UK homes www.energysavingtrust.org.uk/scotland/organisations/ national-and-local-government-HEED
International Energy Agency (IEA): energy efficiency	Global energy statistics and publications on energy efficiency in all sectors, including buildings and appliances www.iea.org/topics/energyefficiency
Inventory of Carbon and Energy (ICE) database	Developed by University of Bath, now available on Circular Economy website. Open access Excel spreadsheet www.circularecology.com/embodied-energy-and-carbon-footprint-database.html
Nicola Terry blog	Provides research and discussions of issues associated with buildings' energy demand http://nicola.qeng-ho.org
Odyssee database	Database on energy efficiency data and indicators for the EU-28 + Norway www.odyssee-mure.eu/
RETScreen 4 (Natural Resources Canada)	Open access project evaluation Excel software tool for renewable energy, energy efficiency and energy performance analysis www.retscreen.net/ang/home.php
US Energy Information Administration	Annual and monthly US energy data including energy demand by sector www.eia.gov/totalenergy/data/annual/index.cfm
WRAP and the UK Green Building Council embodied carbon database	Open access database of embodied carbon data for buildings www.wrap.org.uk/content/embodied-carbon-database. Video of database: www.youtube.com/watch?v=bZt-kP8PIoA

Organisations and professional bodies

Association for the Conservation of Energy (ACE)	Aims to encourage national awareness of the need for and benefits of energy conservation, to establish a sensible and consistent national policy and programme, and to increase investment in energy-saving measures in the UK and Europe www.ukace.org
Association for Environmental Conscious Building (AECB)	Independent not-for-profit organisation that promotes excellence in design and construction www.aecb.net
Building Research Establishment (BRE)	UK organisation that provides advice on realising better buildings, communities and businesses www.bre.co.uk

Carbon Trust	UK organisation that advises businesses and other organisations on reducing their energy, water and carbon www.carbontrust.com/home
Centre for Sustainable Energy (CSE)	Independent charity giving advice, managing innovative energy projects, training and supporting others to act, and undertaking research and policy analysis www.cse.org.uk
Committee on Climate Change (CCC)	Independent UK statutory body whose purpose is to report to Parliament on progress made in reducing greenhouse gas emissions, prepare for climate change and advise on emissions targets www.theccc.org.uk
Energy Saving Trust	UK organisation that provides information and undertakes research aimed at saving energy in homes, businesses and other organisations www.energysavingtrust.org.uk
Global Buildings Performance Network	Globally organised and regionally focused organisation that aims to provide policy expertise and technical assistance to improve building energy performance and realise sustainable built environments for all www.gbpn.org
Innovate UK – formally Technology Strategy Board	UK fund, supports and connects innovative businesses to accelerate sustainable economic growth www.gov.uk/government/organisations/innovate-uk
International Energy Agency's Energy in Buildings and Communities Programme	International energy research and innovation programme in the buildings and communities field. Provides scientific reports and summary information for policy-makers www.iea-ebc.org
Institute of Civil Engineers (ICE)	Professional organisation that supports and promotes civil engineering www.ice.org.uk
National Energy Action	Aims to eradicate fuel poverty by improving and promoting energy efficiency. Campaigns for greater investment in energy efficiency to help those who are poor and vulnerable. www.nea.org.uk
Soft Landings (BSIRIA)	Describes the background and procedures for the Soft Landings approach www.softlandings.org.uk
UK Energy Research Centre (UKERC)	A focal point for UK energy research and a gateway to international energy research communities undertaking interdisciplinary, whole-systems research www.ukerc.ac.uk
US Department of Energy	Range of information on energy-efficient homes, retrofitting, etc. http://energy.gov/energysaver/energy-saver

EU and UK policy documents

EC Commission: buildings	http://ec.europa.eu/energy/en/topics/energy-efficiency/buildings
EU Directive 2010/31/EU (EPBD recast 2010)	http://eur-lex.europa.eu/legal-content/EN/ALL/?uri=OJ:L:2010:153:TOC
UK Carbon Plan	www.gov.uk/government/publications/the-carbon-plan-reducing-greenhouse-gas-emissions--2
UK Climate Change Act 2008	www.legislation.gov.uk/ukpga/2008/27/contents
UK Energy Efficiency Strategy	www.gov.uk/government/collections/energy-efficiency-strategy

Lifecycle assessment

EU Platform on Life Cycle Assessment	http://eplca.jrc.ec.europa.eu
European Reference Life-cycle Database	http://eplca.jrc.ec.europa.eu/ELCD3
Life Cycle Initiative	Initiative between UNEP and SETAC that aims to promote lifecycle thinking, facilitate knowledge exchange and enable users to put lifecycle thinking into practice www.lifecycleinitiative.org
Society of Environmental Toxicology and Chemistry (SETAC)	Supports and promotes LCA research www.setac.org
US Life Cycle Inventory Database	www.nrel.gov/lci

Passivhaus standard

Denby Dale	Passivhaus blog from Project Manager (Bill Butcher) covering a range of design and construction issues www.greenbuildingstore.co.uk/page--passivhaus-diaries.html
Technical film briefing	www.greenbuildingstore.co.uk/page--denby-dale-passivhaus-technical-film.html
E-book on Passivhaus interrogating the technical, economic and cultural challenges of delivering the PassivHaus standard in the UK	Each chapter is a UK Passivhaus case study https://kar.kent.ac.uk/44559/1/PassivHaus_UK_eBook.pdf

Passivhaus Trust UK	Leading UK Passivhaus organisation www.passivhaustrust.org.uk
Passive House Institute	http://passiv.de/en
Passivhaus International (iPHA)	www.passivehouse-international.org
Passipedia	http://passipedia.passiv.de/passipedia_en/start
Potwine Passive House Blog	Blog revealing the underlying rationale and construction approach http://potwinepassive.blogspot.co.uk

Journals

Applied Energy
Architectural Science and Review
Buildings
Building and Environment
Buildings Research and Information
Energy
Energy and Buildings
Energy Efficiency
Energy Research and Social Science
Energy Policy

Recent journal special issues

Journal	Special edition title	Website
Building and Environment (2015) 88, 1–150	Interactions between humans and the built environment	www.sciencedirect.com/science/journal/03601323/88
Building Research & Information (2015) 43(3)	Counting the costs of comfort	www.tandfonline.com/toc/rbri20/current#.VSfsgfnF_To
Building Research & Information (2014) 42(4)	Energy retrofits of owner occupied homes	www.tandfonline.com/toc/rbri20/42/4
Building and Environment (2012) 55, 1186	The implications of a changing climate for buildings	www.sciencedirect.com/science/journal/03601323/55
Buildings (2014) 4(4)	Low carbon building design	www.mdpi.com/journal/buildings/special_issues/low-carbon-build#published
Energy and Buildings (2012) 46, 1–176.	Sustainable and healthy buildings	www.sciencedirect.com/science/journal/03787788/46

Journal	Special edition title	Website
Energy and Buildings (2014) 68(B) 633–720	The 2nd International Conference on Building Energy and Environment 2012	www.sciencedirect.com/ science/journal/03787788/ 68/part/PB
Intelligent Buildings International (2015) 7(2–3)	Designing intelligent school buildings: what do we know	www.tandfonline.com/toc/ tibi20/current#.VR8LY vnF9Zs
Intelligent Buildings International 2013 5(3), 133–196	Post-occupancy evaluation	www.tandfonline.com/toc/ tibi20/5/3#.VR8P9_nF9Zs
Technology Analysis & Strategic Management (2014) 26(10)	Smart metering technology and society	www.tandfonline.com/toc/ ctas20/26/10#.VR8Izvn F9Zt

Index